ISABEL THOMAS

# SELF-DESTRUCTING SCIENCE SPACE

Illustrated by NIKALAS CATLOW



## COUNTDOWN TO SELF-DESTRUCTION

Ever dreamed of exploring space?
What do you think you'd need?

### CHECKLIST

- ✗ Spacesuit
- ✗ Rocket
- ✗ Years of training
- ✗ Planet-sized brain
- ✓ THIS BOOK AND A PAIR OF SCISSORS

Yup, your mission to space starts now! This book is packed with space activities that use every page in a different way.

It's not one of those Zzzzzzzz activity books that's all about sitting quietly and doodling. This book goes off with a BANG!

Get ready to tear, fold, cut, construct and experiment... but be warned:

**THIS BOOK WILL SELF-DESTRUCT!**

BLOOMSBURY
Activity Books

Cut out these cards to fit the universe into a year on page 11.
First mammals
First stars and galaxies
First dinosaurs
First life on Earth
The Milky Way takes shape
First modern humans
The Sun forms
Earth forms
First plants on land
This book self destructs
The Moon forms
First cities
Big Bang
BOOM
Dinosaurs become extinct

# HOW TO ~~USE~~ WRECK THIS BOOK

Space is a big place to explore. Where will you start?

## TRAINING CAMP

Cut along solid lines

Fold along dashed lines

WARNING! THIS PAGE WILL SELF-DESTRUCT!

Look out for this icon – these pages will completely self-destruct, so make sure you read every word before getting started!

## SAFETY INFORMATION FOR MISSION CONTROL

All the projects in this book have been designed to be safe for children to carry out at home, if they are confident with using scissors. We recommend that an adult helps with or supervises certain projects, including Gravity race (page 12), Keeping time (page 25), Don't get dazzled (page 22), and Pocket rocket (page 17). You should never look at the sun directly or with any type of binoculars or telescope. When launching the Pocket rocket (page 17), make sure that it is pointed away from other people.

### Extra kit

Most projects just use scissors, glue and colouring pencils. These boxes tell you if you need to pack any extra kit.

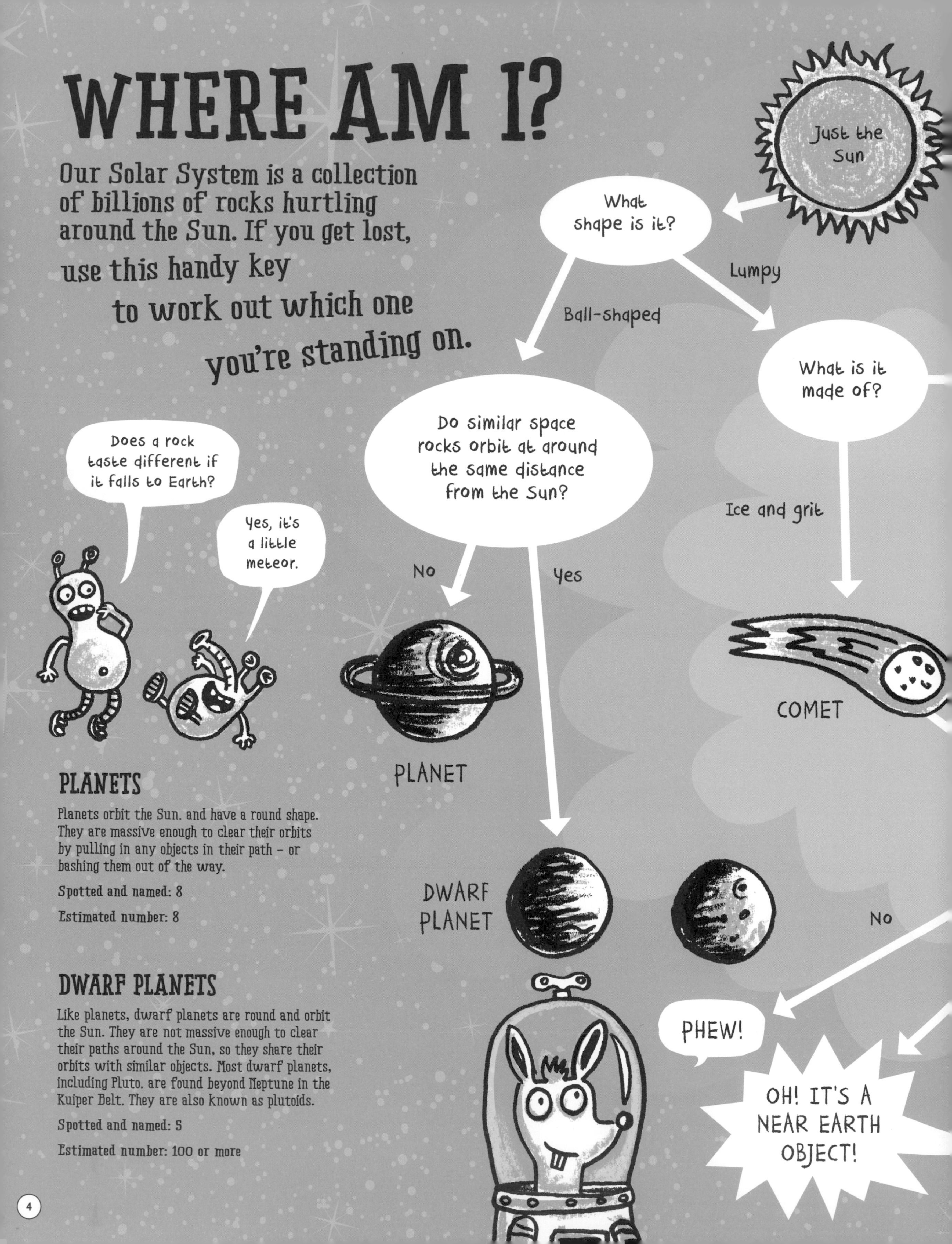


# WHERE AM I?

Our Solar System is a collection of billions of rocks hurtling around the Sun. If you get lost, use this handy key to work out which one you're standing on.

## PLANETS

Planets orbit the Sun. and have a round shape. They are massive enough to clear their orbits by pulling in any objects in their path – or bashing them out of the way.

Spotted and named: 8

Estimated number: 8

## DWARF PLANETS

Like planets, dwarf planets are round and orbit the Sun. They are not massive enough to clear their paths around the Sun, so they share their orbits with similar objects. Most dwarf planets, including Pluto. are found beyond Neptune in the Kuiper Belt. They are also known as plutoids.

Spotted and named: 5

Estimated number: 100 or more

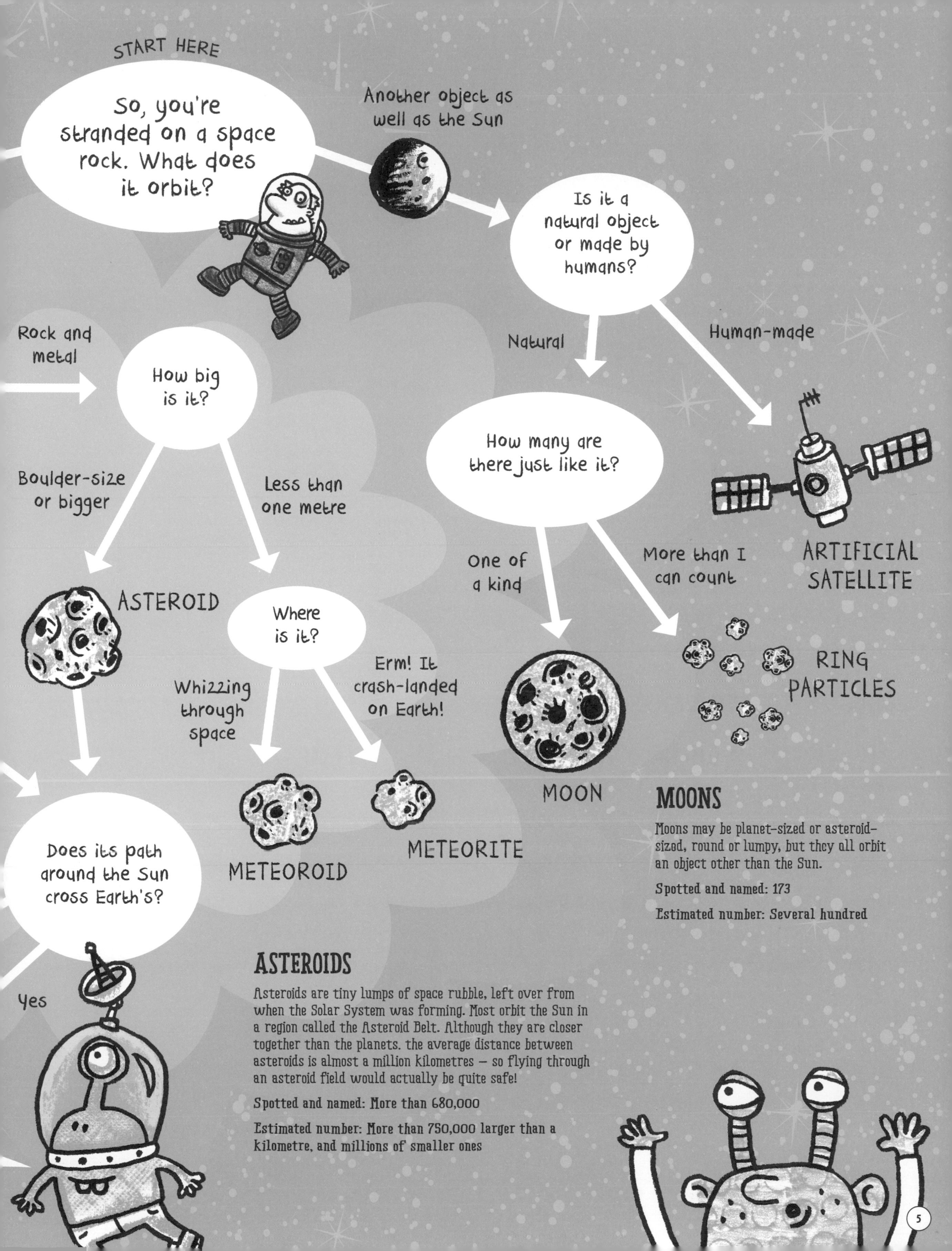


## MOONS

Moons may be planet-sized or asteroid-sized, round or lumpy, but they all orbit an object other than the Sun.

Spotted and named: 173

Estimated number: Several hundred

## ASTEROIDS

Asteroids are tiny lumps of space rubble, left over from when the Solar System was forming. Most orbit the Sun in a region called the Asteroid Belt. Although they are closer together than the planets, the average distance between asteroids is almost a million kilometres – so flying through an asteroid field would actually be quite safe!

Spotted and named: More than 680,000

Estimated number: More than 750,000 larger than a kilometre, and millions of smaller ones

# THE BIG BANG(ER)

The universe is EVERYTHING there is – from giant galaxies to miniscule mites. Scientists have worked out that it all started with a huge explosion known as the Big Bang. Can you make a big bang of your own, using only this page?

The biggest banger ever folded like this was almost three metres tall.

Extra kit

- Scissors

What to do:

1. Take this page out of the book. Crease all the white dotted lines firmly and unfold.
2. Fold the four corners in to the matching symbols.
3. Fold the paper in half, so the long sides meet.
4. Bring the top points down so the folded edges meet along the blue line.
5. Turn the paper over and fold in half so the white dots meet.
6. Hold the bottom point between your thumb and forefinger. Raise your arm and move it down as quickly as possible. Listen for a big bang, and watch stars and galaxies explode from the centre

The bigger the banger, the bigger the bang!

Your banger won't work in space. Sound needs something to travel through, but space is completely empty.

## WHAT WAS THE BIG BANG?

Almost 14 billion years ago, everything in the universe was squashed together into a bubble too tiny to see, and trillions of times hotter than the Sun. When the Big Bang happened, the universe began to expand and cool down very quickly. As this happened, space matter began to form and eventually, after about 200 million years, this made the first stars. These huge nuclear reactors created the materials that make up everything else, from planets and moons to living things.

Try making bangers out of different types of paper, to find out which makes the loudest bang. Newspaper and brown parcel paper give great results!

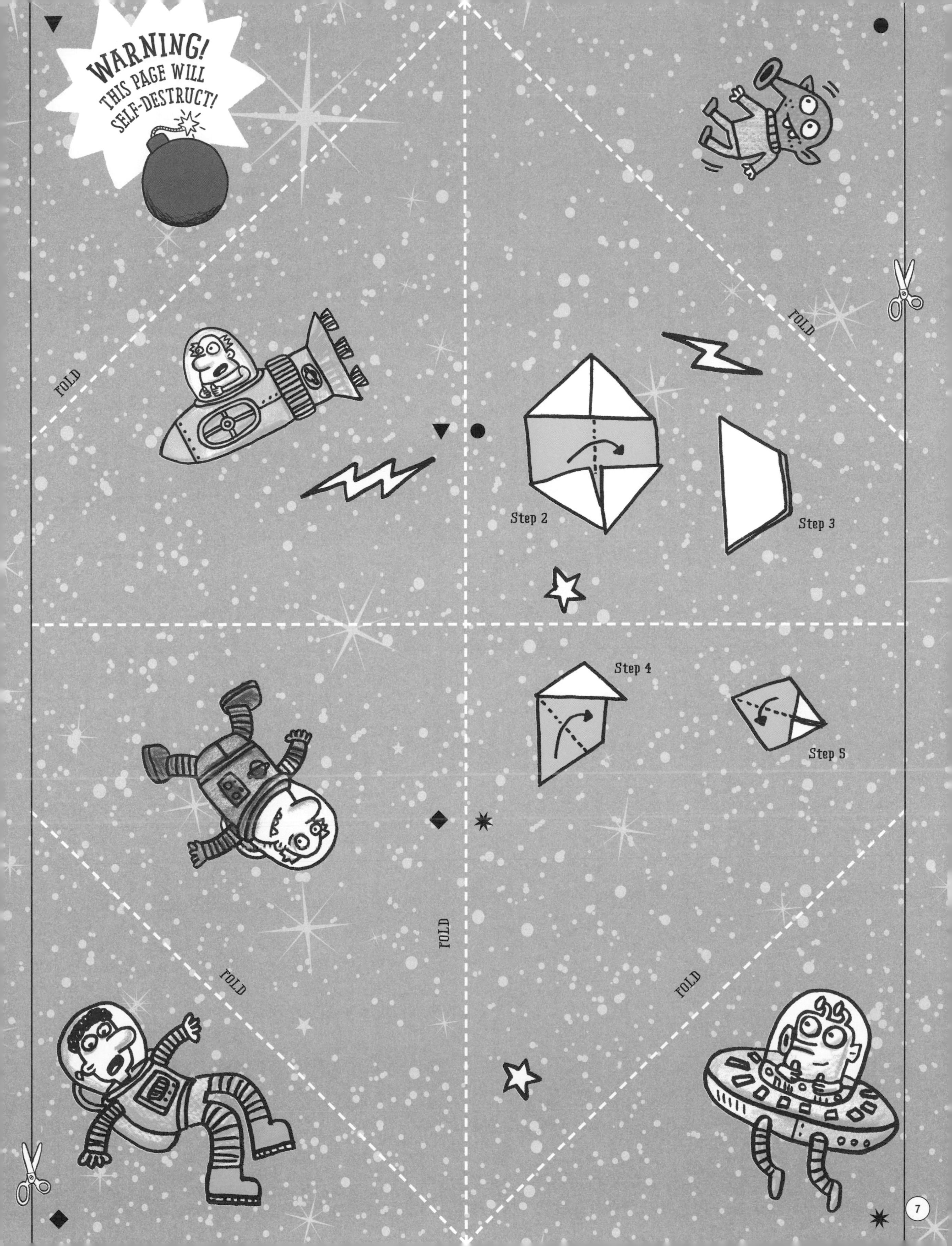
WARNING!
THIS PAGE WILL
SELF-DESTRUCT!
FOLD
FOLD
Step 2
Step 3
Step 4
Step 5
FOLD
FOLD
FOLD

Step 6
FOLD
FOLD
FOLD
FOLD

| o | n | m | l | k | j | i | h | g | f | e | d | c | b | a |
|---|---|---|---|---|---|---|---|---|---|---|---|---|---|---|

# NOT-SO-NEAR NEIGHBOURS

They call it 'space' for a reason... the distances between planets are huge. This page turns into a handy scale map of the Solar System, to help you work out how many sandwiches to pack for the journey!

## Extra kit

- Scissors
- Sticky tape or glue

## What to do

1. Cut this page out of the book. Follow the guides to cut it into 15 strips.
2. Match up the letters to stick the strips together in the right order. Orange tabs with the same letter should overlap.
3. Starting from Neptune, roll the strip in a coil around a pencil. Get your sandwiches ready and unroll the paper to start your journey through the solar system.

WARNING! THIS PAGE WILL SELF-DESTRUCT!

The scale of your map will be one millimetre = one million kilometres!

Here are some things to think about on your journey. How do the distances between planets change as you travel through the solar system? How big would the Sun look from each planet? Are those alien spaceships between Mars and Jupiter, or just the asteroid belt?

One naughty astronaut really did take a sandwich into space. NASA astronaut John Young snuck a corned beef sandwich into his spacesuit before he went into orbit in 1965. It was a bad idea - when he tried to eat it, crumbs began floating around the cabin threatening to damage the electronics!

Did you know that the planets don't travel in perfect circles around the Sun? They have elliptical orbits (a slightly squashed circle), so their distance from the Sun is always changing.

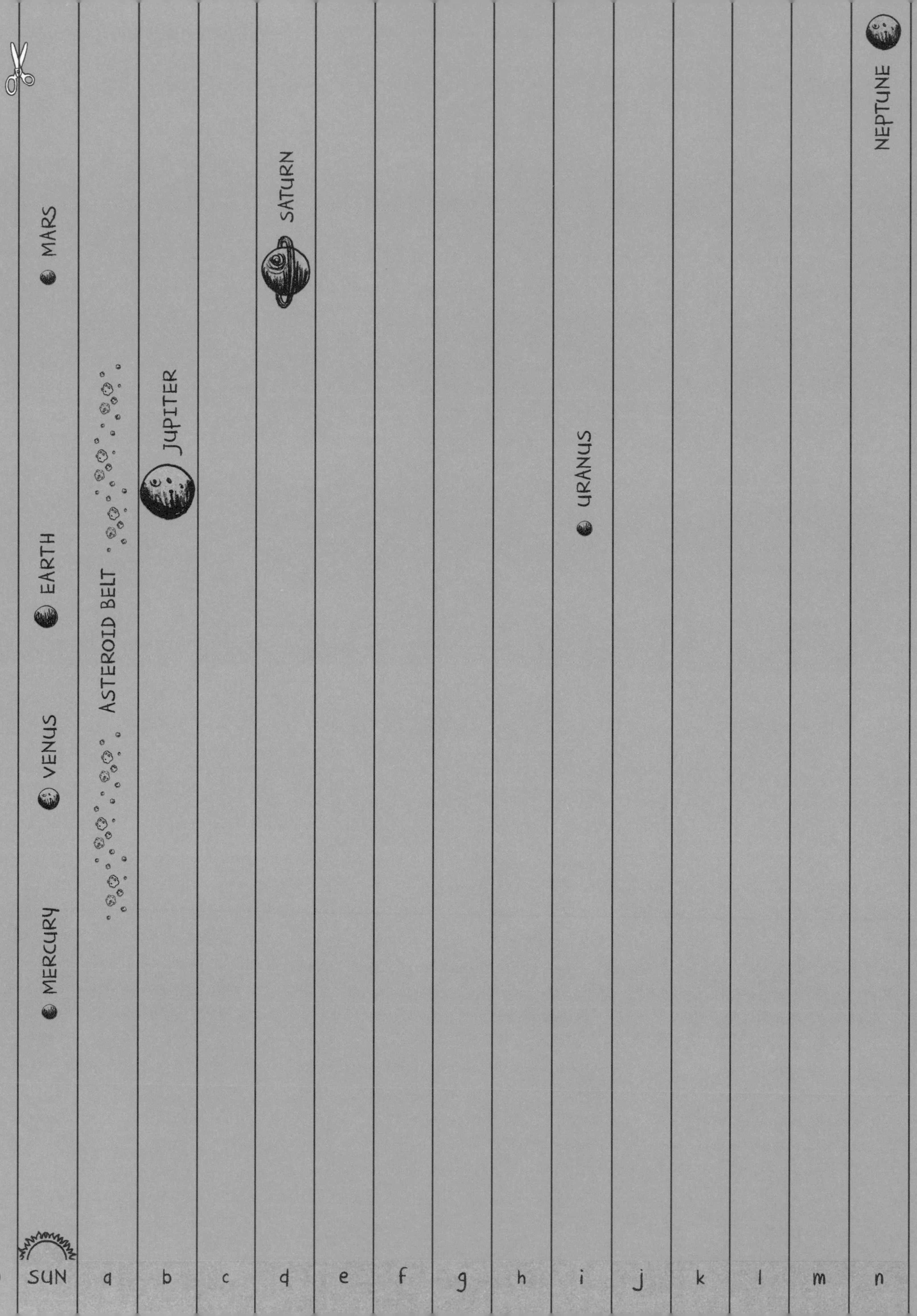

MERCURY
VENUS
EARTH
MARS
ASTEROID BELT
JUPITER
SATURN
URANUS
NEPTUNE

SUN a b c d e f g h i j k l m n

# HOW TO BUILD A UNIVERSE

Imagine squishing the 13.7-billion-year history of the universe into a single year. If the Big Bang happens right at the start, how long does it take for our planet to appear, and fill up with plants, animals and people?

**Extra kit:**

- Universe in a year cards (page 2)
- Scissors
- Sticky tack

**What to do:**

1. Cut out the cards on page 2. Each one is an event in the history of the universe. If the universe was formed in a year, when do you think each event happened? Stick the cards on the calendar, then check the inside of the back cover to see if you were right!

What would be happening on your birthday?

The year begins with the ultimate fireworks display – the Big Bang!

On this timescale, each month represents just over one billion years of cosmic history!

**JANUARY**

1 2 3 4 5 6 7
8 9 10 11 12 13 14
15 16 17 18 19 20 21
22 23 24 25 26 27 28
29 30 31

**FEBRUARY**

1 2 3 4 5 6 7
8 9 10 11 12 13 14
15 16 17 18 19 20 21
22 23 24 25 26 27 28
29

**MARCH**

1 2 3 4 5 6 7
8 9 10 11 12 13 14
15 16 17 18 19 20 21
22 23 24 25 26 27 28
29 30 31

**APRIL**

1 2 3 4 5 6 7
8 9 10 11 12 13 14
15 16 17 18 19 20 21
22 23 24 25 26 27 28
29 30

**MAY**

1 2 3 4 5 6 7
8 9 10 11 12 13 14
15 16 17 18 19 20 21
22 23 24 25 26 27 28
29 30 31

**JUNE**

1 2 3 4 5 6 7
8 9 10 11 12 13 14
15 16 17 18 19 20 21
22 23 24 25 26 27 28
29 30

**JULY**

1 2 3 4 5 6 7
8 9 10 11 12 13 14
15 16 17 18 19 20 21
22 23 24 25 26 27 28
29 30 31

**AUGUST**

1 2 3 4 5 6 7
8 9 10 11 12 13 14
15 16 17 18 19 20 21
22 23 24 25 26 27 28
29 30 31

**SEPTEMBER**

1 2 3 4 5 6 7
8 9 10 11 12 13 14
15 16 17 18 19 20 21
22 23 24 25 26 27 28
29 30

**OCTOBER**

1 2 3 4 5 6 7
8 9 10 11 12 13 14
15 16 17 18 19 20 21
22 23 24 25 26 27 28
29 30 31

**NOVEMBER**

1 2 3 4 5 6 7
8 9 10 11 12 13 14
15 16 17 18 19 20 21
22 23 24 25 26 27 28
29 30

**DECEMBER**

1 2 3 4 5 6 7
8 9 10 11 12 13 14
15 16 17 18 19 20 21
22 23 24 25 26 27 28
29 30 31

# GRAVITY RACE

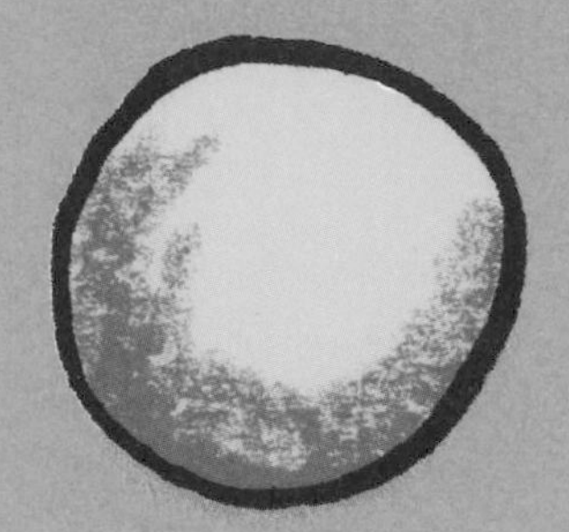

What's faster – this page, or your trainers? With your help, this page can win a gravity race!

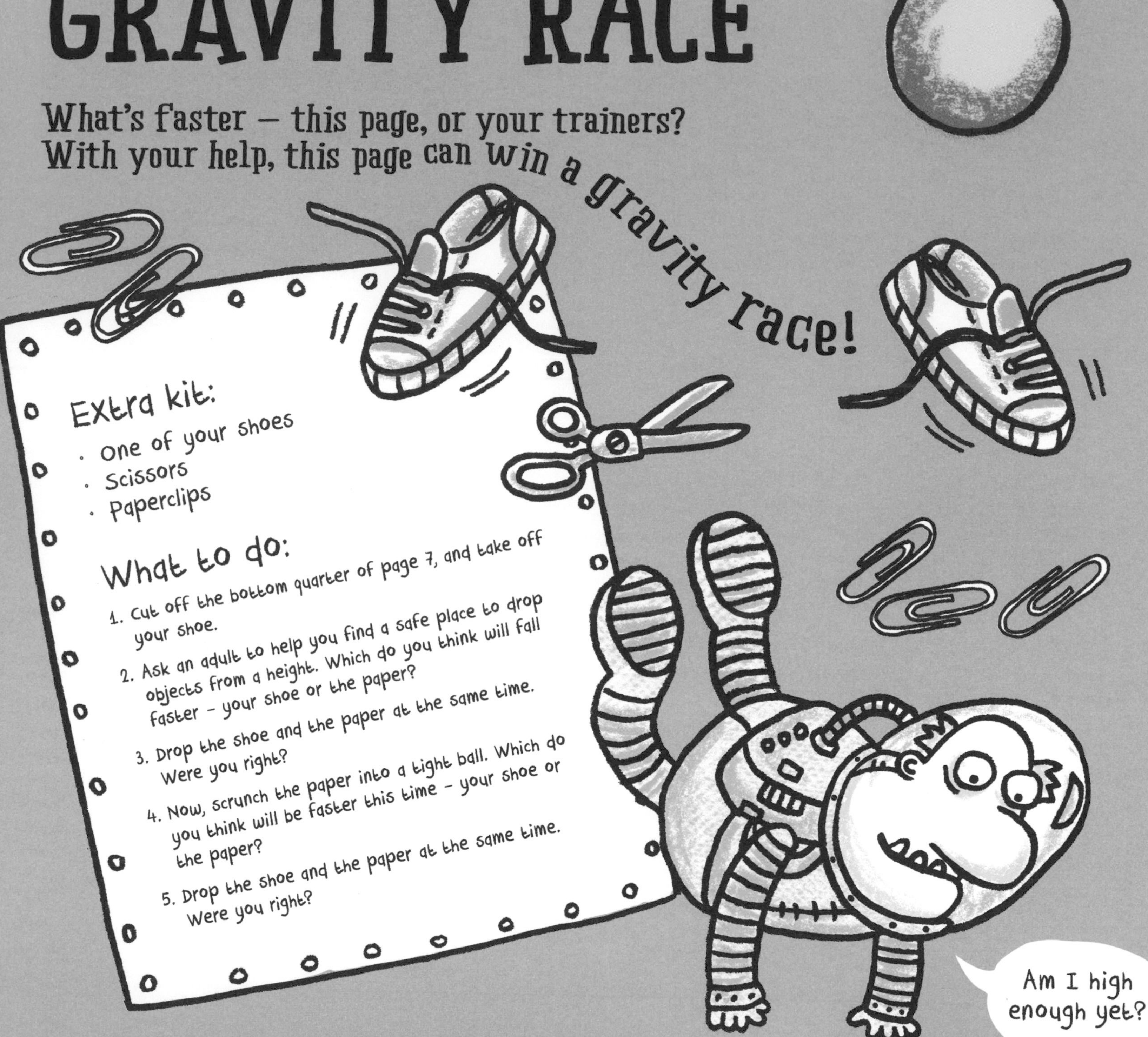

Extra kit:

- One of your shoes
- Scissors
- Paperclips

What to do:

1. Cut off the bottom quarter of page 7, and take off your shoe.
2. Ask an adult to help you find a safe place to drop objects from a height. Which do you think will fall faster – your shoe or the paper?
3. Drop the shoe and the paper at the same time. Were you right?
4. Now, scrunch the paper into a tight ball. Which do you think will be faster this time – your shoe or the paper?
5. Drop the shoe and the paper at the same time. Were you right?



Objects fall to the ground because of gravity. This is the 'glue' that sticks our Solar System together. It's not a sticky, gloopy glue, but an invisible force that pulls objects towards each other. Every object has a gravitational force, but we only notice the pull of massive objects such as planets. No matter where you drop an object on Earth, gravity will pull it towards the centre of the planet.

To work like a scientist, make sure you are not dropping the objects on to someone's head!

This experiment proves that gravity makes all objects accelerate towards Earth at the same rate. Heavier things don't fall faster than lighter things. So what does change the speed of a falling object? Objects rub against the air as they fall. This 'air resistance' is what slows objects down. A flat piece of paper has a greater surface area than a ball of paper, so the air resistance is greater.

NEVER drop a hammer on Earth!

WARNING! THIS PAGE WILL SELF-DESTRUCT!

For a fair test of gravity, you need to drop objects somewhere without air – such as the Moon! An astronaut once dropped a feather and a hammer on the Moon. They watched them hit the ground at the same time.

D

This piece of paper can put air resistance to work. Turn over to find out how.

Use this piece of paper for the gravity race!

What to do:
Put air resistance to work by making a paper helicopter!
1. Cut off the second quarter of page 14.
2. Cut along the solid lines.
3. Fold flaps A and B over flap C and hold them together at the bottom with a paperclip.
4. Fold flaps D and E in opposite directions to make blades.
5. Throw the helicopter into the air and see what happens.
Try dropping the helicopter and a paperclip at the same time. Which falls faster — the lighter paperclip, or the heavier helicopter?
A
D
C
E
B

# COSMIC DUST

## Don't look now but a meteorite just landed in your garden!

Good places to collect micrometeorites
- The middle of a desert
- Antarctica
- Your back garden

### Extra kit

- Sticky tape
- Very strong magnet
- Modelling clay
- Plastic bag
- Jam jar, plant pot or other container

### What to do

1. Choose a calm, dry day. Cut out the circle below and join the cut and dotted lines to make a cone.
2. Stick the strip of paper below the cone in a loop and weigh it down with a lump of modeling clay. Pop the bottom of the cone into a heavy container such as a jam jar, and put it outside. Choose a high spot, away from roads, buildings and trees.
3. Leave the cone for up to 48 hours. Bring it inside without disturbing the dust that has collected in it. Some of this dust is space dust!
4. To identify micrometeorites, pop a strong magnet into a plastic bag. Hold it on the outside of the cone and slowly pull it towards the top. Any dust or grit that travels with it is magnetic and may be a micrometeorite. Let these particles cling to the magnet and repeat.
5. Hold the magnet over a piece of white paper. Remove the magnet from the bag so the magnetic dust falls onto the paper.
6. Use sticky tape to hold the particles in place, and examine them.

TEMPLATE

1.

2.

4.

Every year, more than nine tonnes of micrometeorites plunge into Earth's atmosphere and fall to the ground. They don't leave craters. After a bumpy journey through the atmosphere, most are so micro you could line up six on a human hair. But some are big enough to collect.

WARNING! THIS PAGE WILL SELF-DESTRUCT!

What do you call a space rock that misses Earth?

## DON'T GET COSMIC DUST ON THE CARPET!

These tiny grains of rock were once whizzing around the Solar System as part of a comet or asteroid. Collisions in space send chunks of rock hurtling towards Earth every day. Most burn up to nothing as they fall through the atmosphere, but some micrometeorites make it to the ground, at a rate of one per square metre per day. Every footstep you take outside will pick up grains of cosmic dust!

How did the astronaut keep up his trousers?

With an asteroid belt!

## WHY ARE METEORITES MAGNETIC?

Most space rocks contain a lot of magnetic metal, such as iron and nickel. This helps you tell them apart from other dust and grit in your garden, such as ash or pollen grains. Earth rocks can also be magnetic, so look out for these other signs too.

### METEORITE CHECKLIST

Unless you are very lucky, the meteorites collected in your cone will be less than a millimetre big! But bigger ones do land on Earth. Here's how to tell them apart from normal rocks.

Magnetic

Strangely lumpy

Dark crust

Very heavy for its size

# POCKET ROCKET

This small but speedy rocket has all the right parts – a nose cone, rocket body and fins for a smooth flight. You're just missing four million litres of explosive fuel. But don't worry –

this rocket is launched

with puff power!

A real rocket is this shape – but 500 times bigger!

The body of a real rocket is not empty. It's full of rocket fuel, which is burned very quickly. Exhaust gases burst out of the bottom of the rocket, sending it in the opposite direction.



## Extra kit

- Bendy drinking straw
- Pens or pencils
- Sticky tape or glue
- Scissors

## What to do

1. Cut out the templates on this page.
2. Decorate each one. Don't forget to name your rocket, so it can be identified if it reaches the Moon!
3. Place a pen or pencil along the long edge of the rocket body. Roll the paper around it to make a long cylinder. Secure the paper cylinder with sticky tape or glue, then slide the pen or pencil out.
4. Bend the nose cone template so the straight edges overlap. Hold in place with sticky tape, then join the cone to the rocket body with tape or glue.
5. Join the fins to the rocket by bending the flaps in opposite directions and gluing them to the rocket body at the base.
6. Bend the straw, and slide the rocket onto the long end. Point the rocket towards the Moon and blow through the straw as hard as you can.

What do you think of that café on the Moon?

Great food, but it had no atmosphere!

## WHY ARE ROCKETS SHAPED LIKE ROCKETS?

To leave Earth's atmosphere, a rocket has to beat the pull of gravity AND push a lot of air out of the way. A pointed nose cone pushes through the air more easily. The narrower the rocket, the less air it has to push out of the way. Because they are narrow, rockets also have to be very tall, to fit in all the fuel needed to blast them into space. The fins help the long, thin rocket to fly in the right direction without wobbling from side to side. Try launching your rocket with and without fins to see the difference!

rocket body

Challenge your friends to a rocket race. How high or far can you make the rocket go?

fin

glue

nose cone

glue

fin

glue

fin

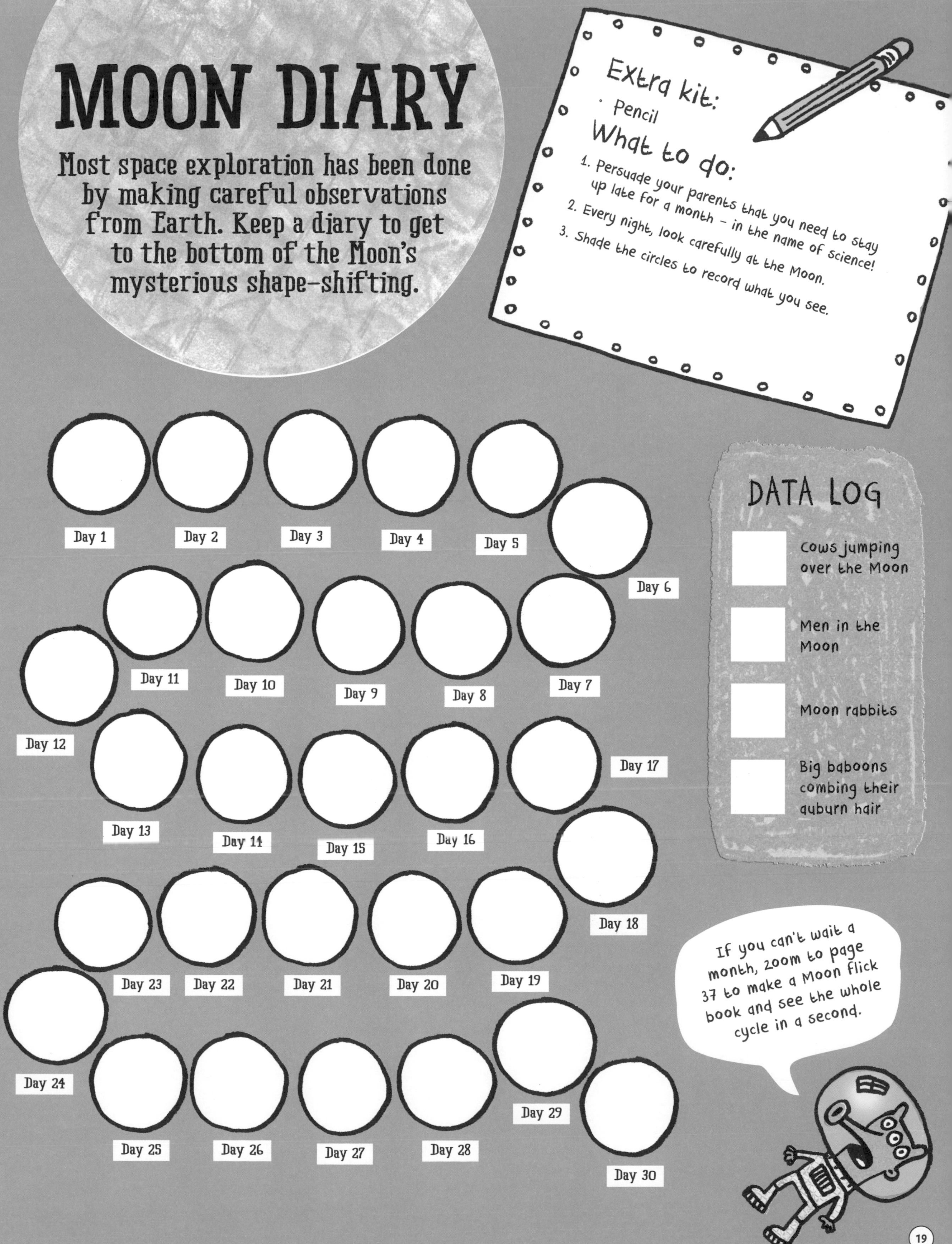
MOON DIARY
Most space exploration has been done by making careful observations from Earth. Keep a diary to get to the bottom of the Moon's mysterious shape-shifting.
Extra kit:
· Pencil
What to do:
1. Persuade your parents that you need to stay up late for a month – in the name of science!
2. Every night, look carefully at the Moon.
3. Shade the circles to record what you see.
Day 1
Day 2
Day 3
Day 4
Day 5
Day 6
Day 7
Day 8
Day 9
Day 10
Day 11
Day 12
Day 13
Day 14
Day 15
Day 16
Day 17
Day 18
Day 19
Day 20
Day 21
Day 22
Day 23
Day 24
Day 25
Day 26
Day 27
Day 28
Day 29
Day 30
DATA LOG
Cows jumping over the Moon
Men in the Moon
Moon rabbits
Big baboons combing their auburn hair
If you can't wait a month, zoom to page 37 to make a Moon flick book and see the whole cycle in a second.

## WHAT'S HAPPENING?

The Moon doesn't really change shape – it's a trick of the light. Just like Earth, one side of the Moon is always bathed in sunlight. 'Moonlight' is really sunlight bouncing off the Moon's surface. As the Moon orbits Earth, we see can see different amounts of the sunlit side. These different views are called phases.

Sometimes the entire sunlit side faces towards us (a full moon). Sometimes we only see part of the sunny side.

The Moon spins as it orbits Earth, so the same side always faces Earth. Until the 1950s, no one knew what was on the other side. Abandoned alien spaceships? Cheese mines? When human spacecraft finally flew around the Moon their pictures showed that the far side of the Moon is... just like the near side!

# NEW MOON

Earth's Moon is close enough for us to get a good look. Over thousands of years, people have imagined seeing different shapes in the dark and light patterns on the surface.

The light and dark areas are actually different types of rock. The light areas are the highlands, the oldest parts of the Moon's surface. The darker rock was formed by lava that flowed into dips and dents on the Moon's surface. These areas are known as 'seas'.

Face

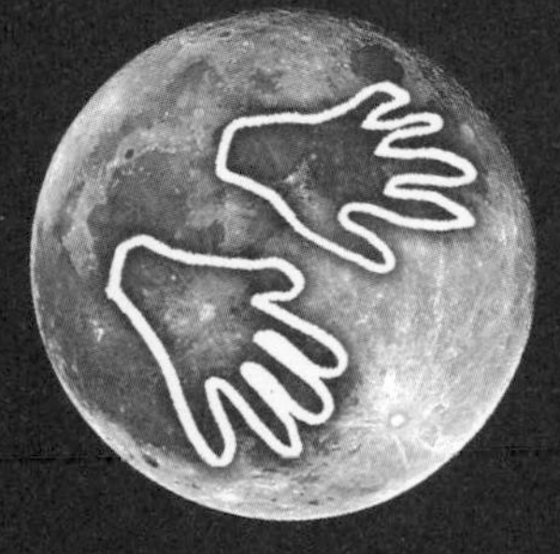
Hands

Rabbit

Frog

What shapes can you see in the Moon? Record them here.

Why didn't you park on the Moon?

It was full.

# DON'T GET DAZZLED

The Sun is a star, but it's much closer to Earth than the stars you can see at night – close enough to damage your eyes if you look at it. This pinhole viewer will help you take a safe look at the Sun.

**NEVER** look at the sun directly, or through any kind of binoculars or telescope. Sunlight can damage your eyes forever, and even cause blindness.

Extra kit:

- Scissors
- Glue or sticky tape
- Drawing pin

What to do:

1. Cut out the template on the next page and fold along the dashed lines. Cut out the rectangular hole on one side.
2. Use a drawing pin to make a small hole at the centre of the cross.
3. Fold the paper into a tube, gluing or taping the tabs in place. Make sure there are no small gaps at the ends.
4. Follow the instructions on the side of the viewer to see the Sun safely.

## WHAT AM I LOOKING AT?

A 4.6-billion-year-old star! Like all stars, the Sun is a huge ball of superheated gas. Nuclear reactions at the centre make huge amounts of energy. This energy escapes from the Sun's surface and zooms across space, hitting Earth eight minutes later!

## SUN SECRETS

Satellites can get a close look the Sun without hurting their eyes. They have recorded enormous explosions on the surface, bubbles of plasma that erupt into space, and solar quakes that ripple across the surface.

WARNING! THIS PAGE WILL SELF-DESTRUCT!

## NOW YOU SEE IT, NOW YOU DON'T

The Sun is 400 times bigger than the Moon. Because the Sun is almost 400 times further away, they look about the same size from Earth. This means that when a full Moon gets between Earth and the Sun, it blocks the Sun's light and causes an eclipse.

TOTAL ECLIPSE – The Moon is perfectly in line with the Sun. It covers the whole Sun and the sky gets dark.

ANNULAR ECLIPSE – The Moon covers the centre of the Sun, but a 'ring of fire' is visible around the edge.

PARTIAL ECLIPSE – The Moon only covers part of the Sun.

How does the Sun keep his hat on?
Eclipse it!
1. Stand with your back to the Sun.
2. Hold the viewer near your forehead, with the pinhole pointing back over your head towards the Sun.
3. Look through the rectangular viewing hole. Adjust the position of the tube until you can see an image of the Sun on the end of the tube opposite the pinhole.
WARNING: Never look at the Sun directly, or through the pinhole.
A longer tube will give you a larger, clearer image of the Sun. Once you have got the hang of making a pinhole viewer, try making a two-metre-long version using cardboard boxes!

Push pencil through centre

6pm 6am 5pm 7am 4pm 8am 3pm 9am 2pm 10am 1pm 11am 12 noon

# KEEPING TIME

As our planet spins in space, the Sun seems to move across the sky. The direction of shadows cast by the Sun changes during the day, giving us a really handy way to tell the time!

WARNING! THIS PAGE WILL SELF-DESTRUCT!

## HOW FAST AM I MOVING WHEN I'M SITTING STILL?

The Earth spins 15° every hour (this is why the hour lines on your sundial are 15° apart). This means that the land at the equator is moving at 1,600 kilometres per hour. At the same time, the planet is zooming around the Sun at just over 100,000 kilometres per hour.

Next time someone asks you to sit still, explain that it's impossible!

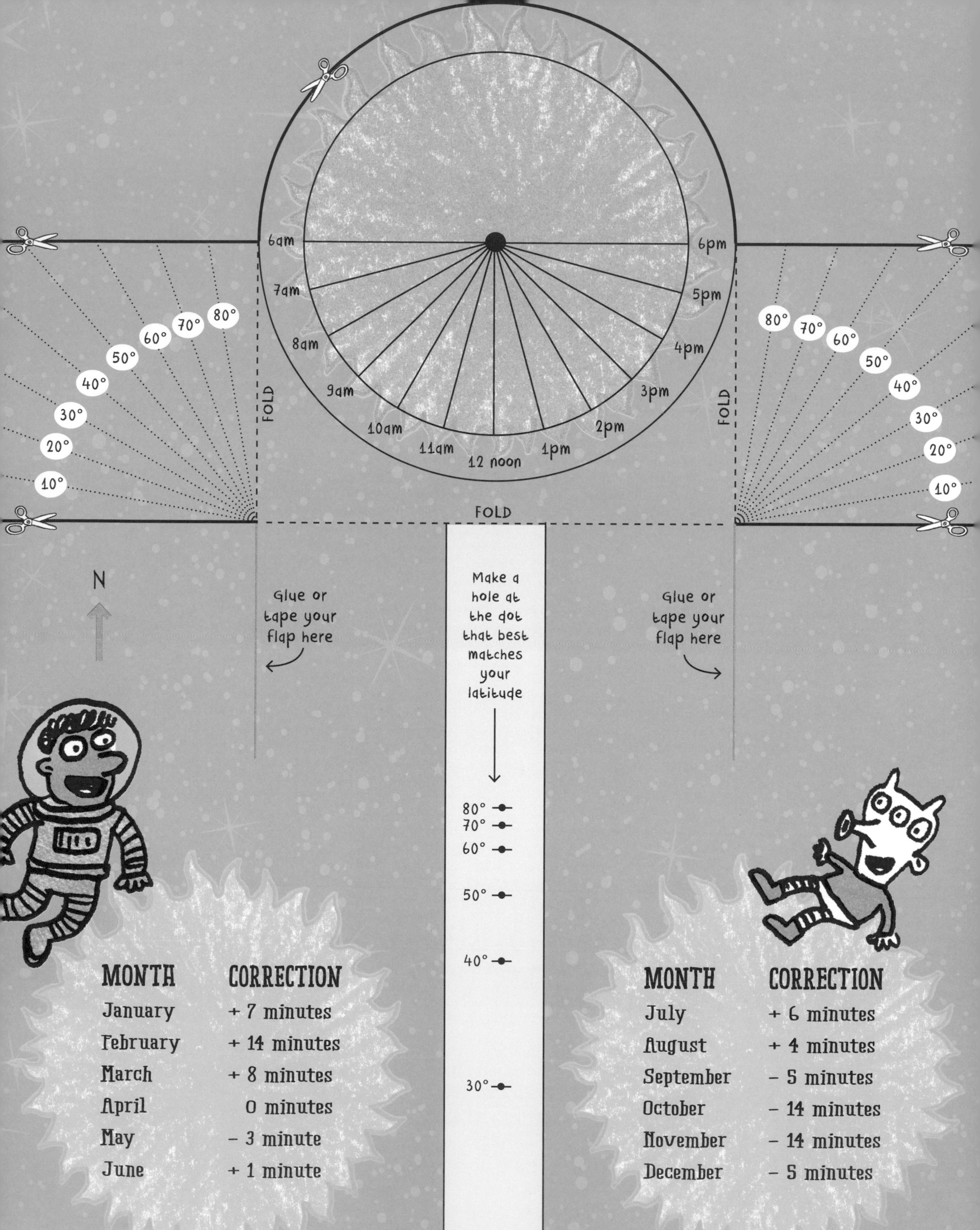


| MONTH | CORRECTION |
|---|---|
| January | + 7 minutes |
| February | + 14 minutes |
| March | + 8 minutes |
| April | 0 minutes |
| May | – 3 minute |
| June | + 1 minute |

| MONTH | CORRECTION |
|---|---|
| July | + 6 minutes |
| August | + 4 minutes |
| September | – 5 minutes |
| October | – 14 minutes |
| November | – 14 minutes |
| December | – 5 minutes |

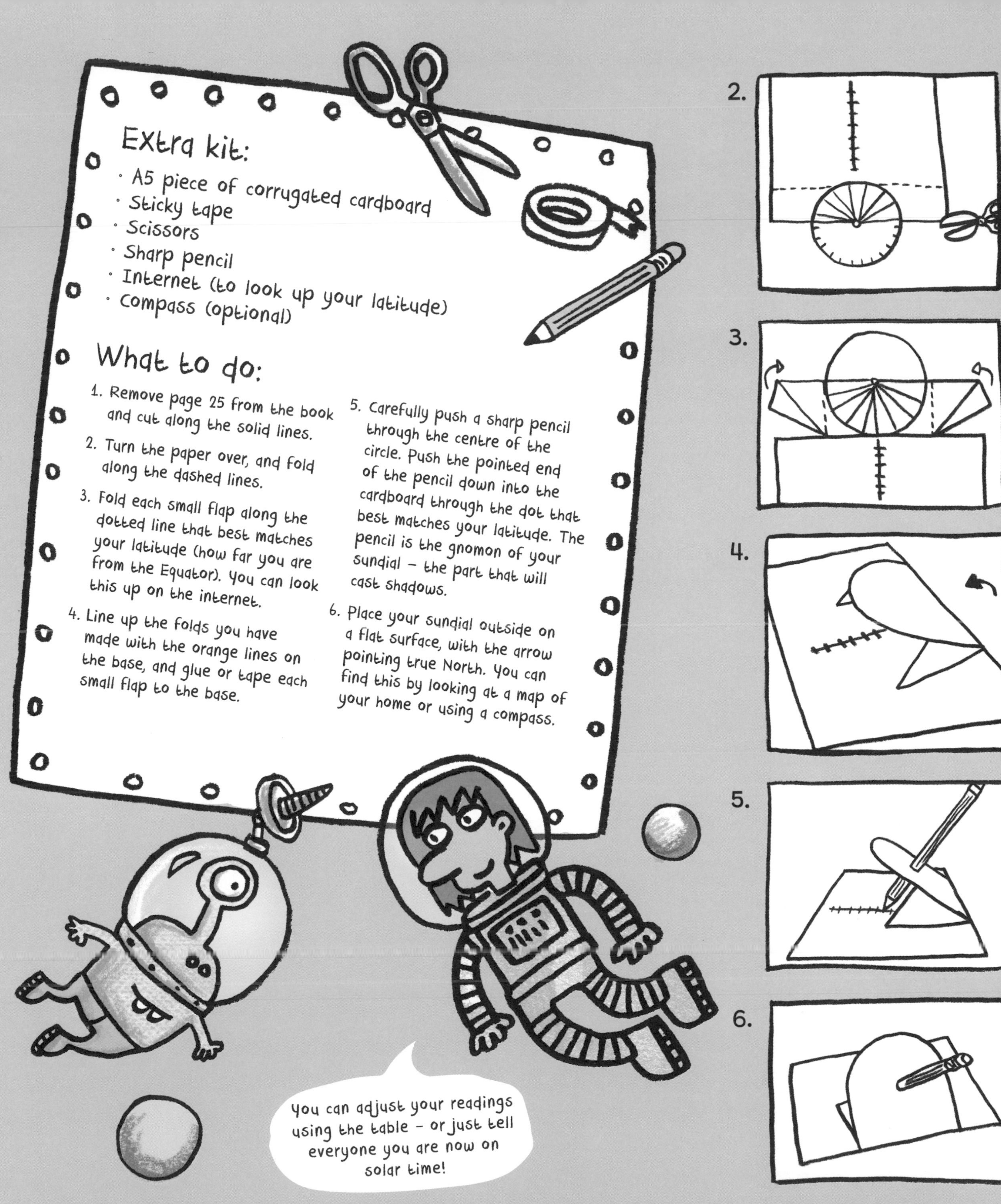

## Extra kit:

- A5 piece of corrugated cardboard
- Sticky tape
- Scissors
- Sharp pencil
- Internet (to look up your latitude)
- Compass (optional)

## What to do:

1. Remove page 25 from the book and cut along the solid lines.
2. Turn the paper over, and fold along the dashed lines.
3. Fold each small flap along the dotted line that best matches your latitude (how far you are from the Equator). You can look this up on the internet.
4. Line up the folds you have made with the orange lines on the base, and glue or tape each small flap to the base.
5. Carefully push a sharp pencil through the centre of the circle. Push the pointed end of the pencil down into the cardboard through the dot that best matches your latitude. The pencil is the gnomon of your sundial – the part that will cast shadows.
6. Place your sundial outside on a flat surface, with the arrow pointing true North. You can find this by looking at a map of your home or using a compass.



2.

3.

4.

5.

6.

The sundial measures 'solar time', or the position of the Sun in the sky. 'Clock time' divides the world into artificial time zones to make life easier. Unless you live right at the centre of your time zone, your sundial reading will be slightly different from clock time. The Sun's position in the sky also changes with the seasons, which affects the sundial reading by up to 15 minutes.

# GET READY FOR YOUR SPACEWALK!

Spacesuits don't just keep astronauts warm. They are life support systems, communications modules and personal jetpacks rolled into one! How quickly can you get this astronaut dressed for her spacewalk?

## Extra kit

- Scissors

## What to do

1. Cut out the astronaut and the spacesuit layers.
2. Follow the checklist to dress your astronaut for space, folding the tabs around the astronaut.

CHECKLIST

- [x] Maximum Absorption Garment
- [x] Liquid Cooling and Ventilation Garment
- [ ] Lower Torso
- [ ] In-Suit Drink Bag
- [ ] Hard Upper Torso
- [ ] Arms
- [ ] Displays and Controls Module (front)
- [ ] Primary Life Support System (back)
- [ ] Communications Carrier Assembly
- [ ] Food Bar
- [ ] Helmet
- [ ] Visor Assembly
- [ ] Gloves
- [ ] Tethers

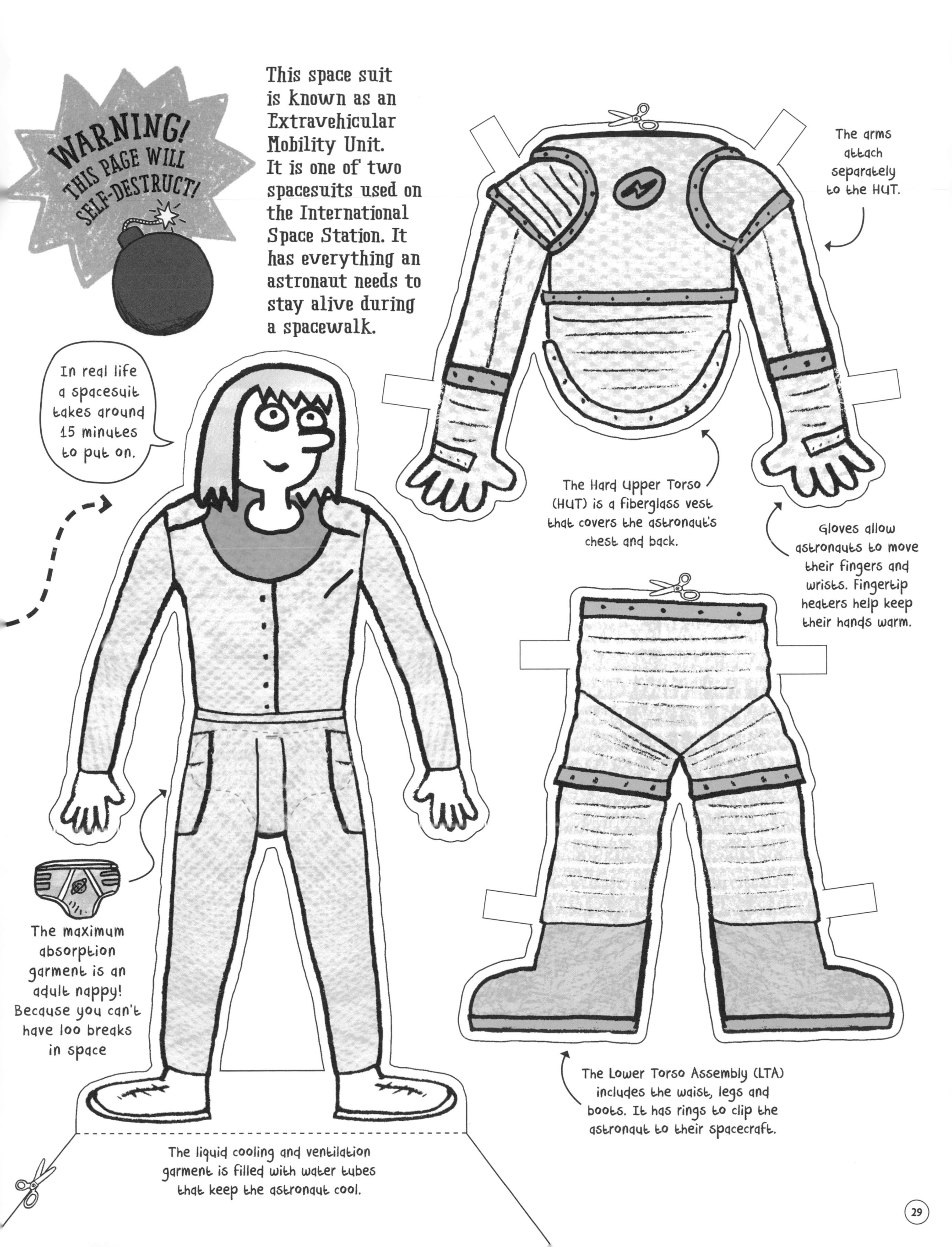
WARNING! THIS PAGE WILL SELF-DESTRUCT!
This space suit is known as an Extravehicular Mobility Unit. It is one of two spacesuits used on the International Space Station. It has everything an astronaut needs to stay alive during a spacewalk.
The arms attach separately to the HUT.
In real life a spacesuit takes around 15 minutes to put on.
The Hard Upper Torso (HUT) is a fiberglass vest that covers the astronaut's chest and back.
Gloves allow astronauts to move their fingers and wrists. Fingertip heaters help keep their hands warm.
The maximum absorption garment is an adult nappy! Because you can't have loo breaks in space
The Lower Torso Assembly (LTA) includes the waist, legs and boots. It has rings to clip the astronaut to their spacecraft.
The liquid cooling and ventilation garment is filled with water tubes that keep the astronaut cool.

In direct sunlight, the temperature in space can be more than fifty seven degrees Celsius. White is the best colour for reflecting heat.

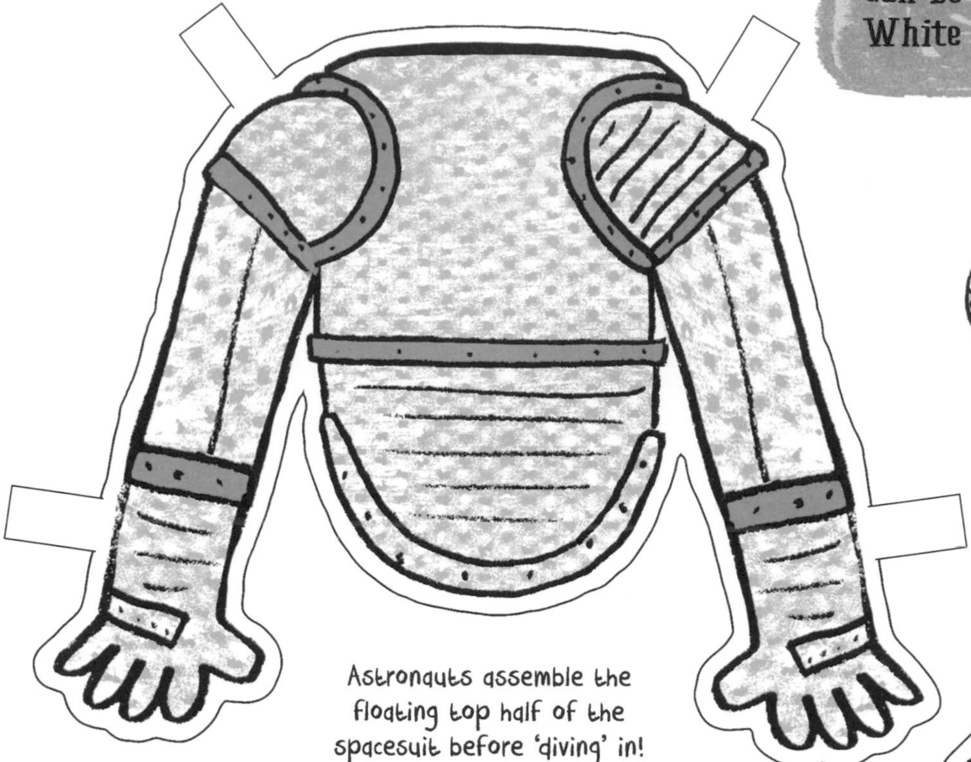

On Earth the spacesuit weighs as much as a baby elephant. Luckily, in space it's weightless!

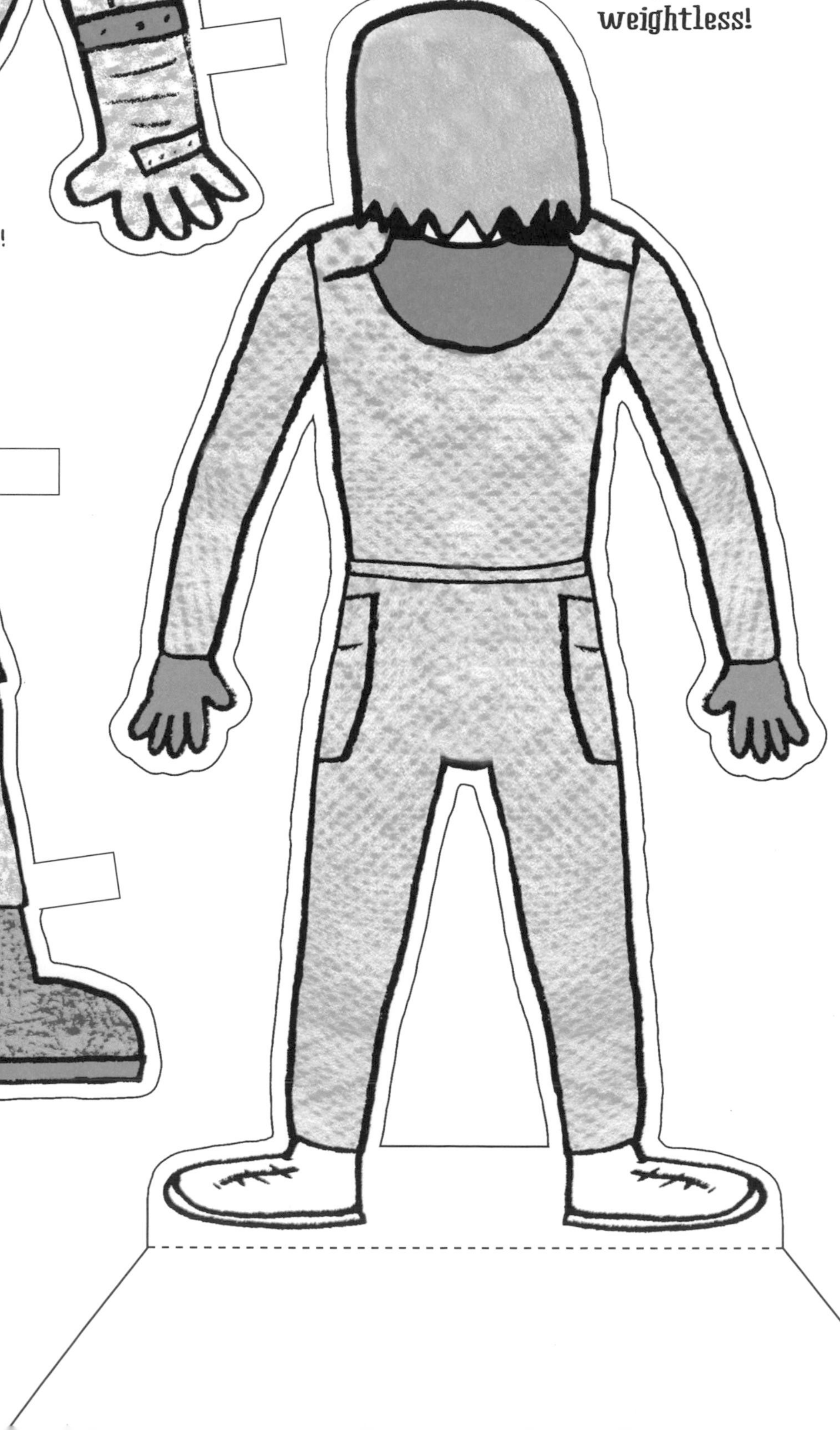

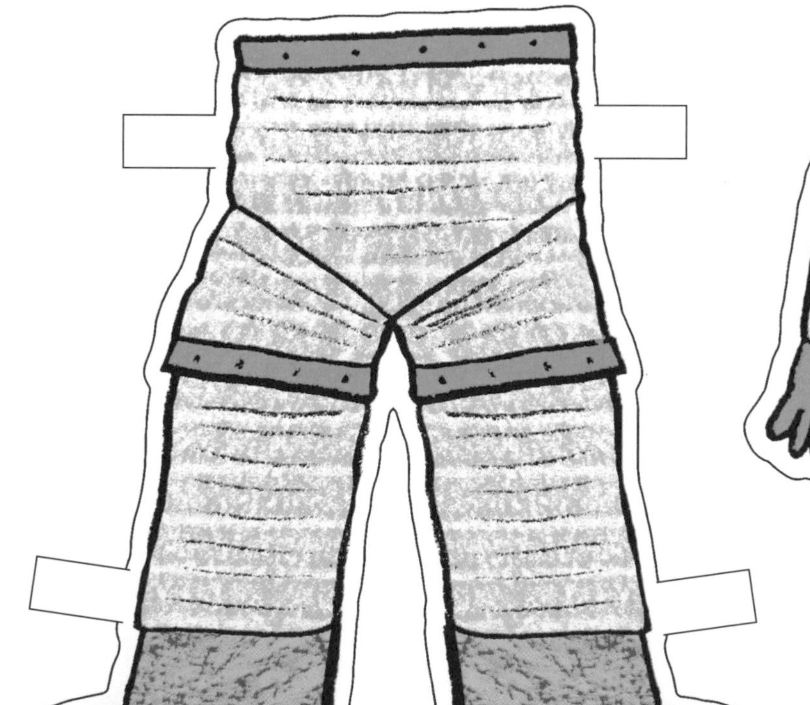

Astronauts assemble the floating top half of the spacesuit before 'diving' in!

The spacesuit has 18,000 parts!

The helmet is a plastic bubble, which seals to the neck of the HUT.
The Communications Carrier Assembly (CCA) is a cap with earphones and microphones built in.
The Visor assembly has three different visors. Headlamps and cameras can also be attached
The Displays and Control Module is mounted on the astronaut's chest.
The Primary Life Support System (PLSS) is worn on the back, like a rucksack. It has breathing equipment, a battery, water-cooling equipment, a fan, a two-way radio, and a warning system.
DON'T FORGET THE ACCESSORIES!
The cuff checklist has a list of the tasks for a spacewalk.
A wrist mirror lets the astronaut see the controls on their chest.
Safety tethers attach the astronaut to the spacecraft. It's just not cool to float off...
This drink bag clips inside the HUT.

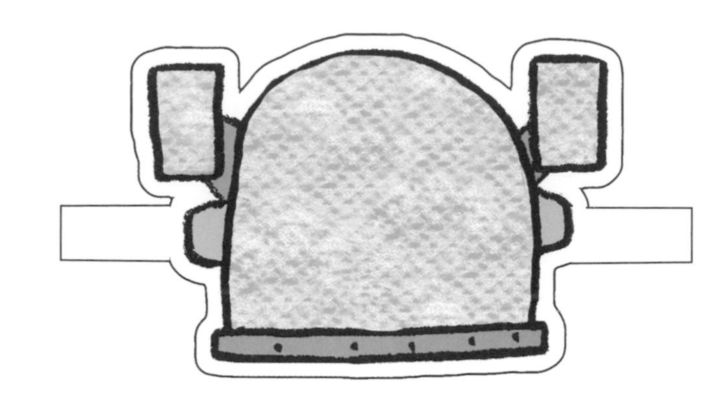

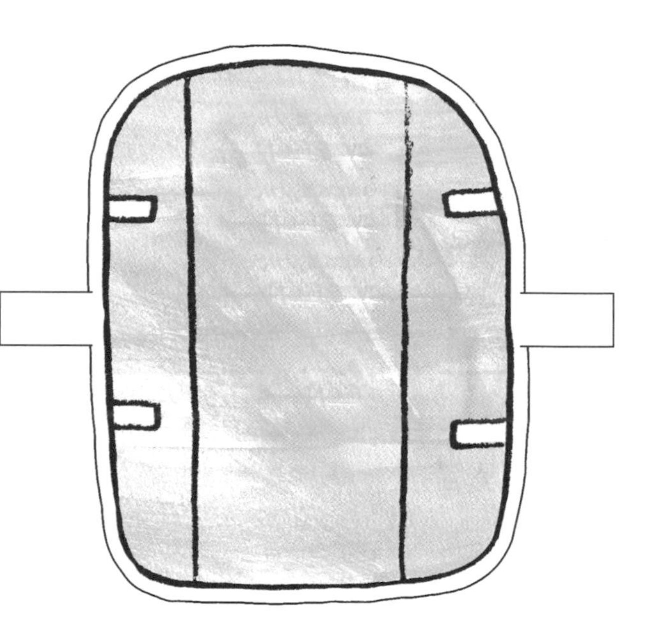

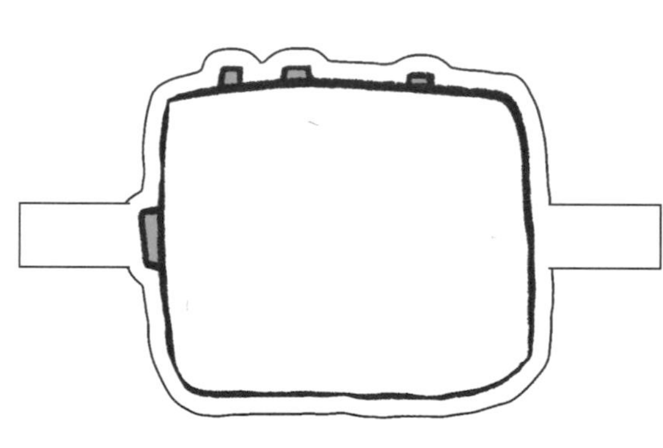

# ROVER RACE

Robot rovers are exploring Mars right now, helping us find out more about this rocky red planet. Can you guide a rover across the Martian landscape without getting stuck?

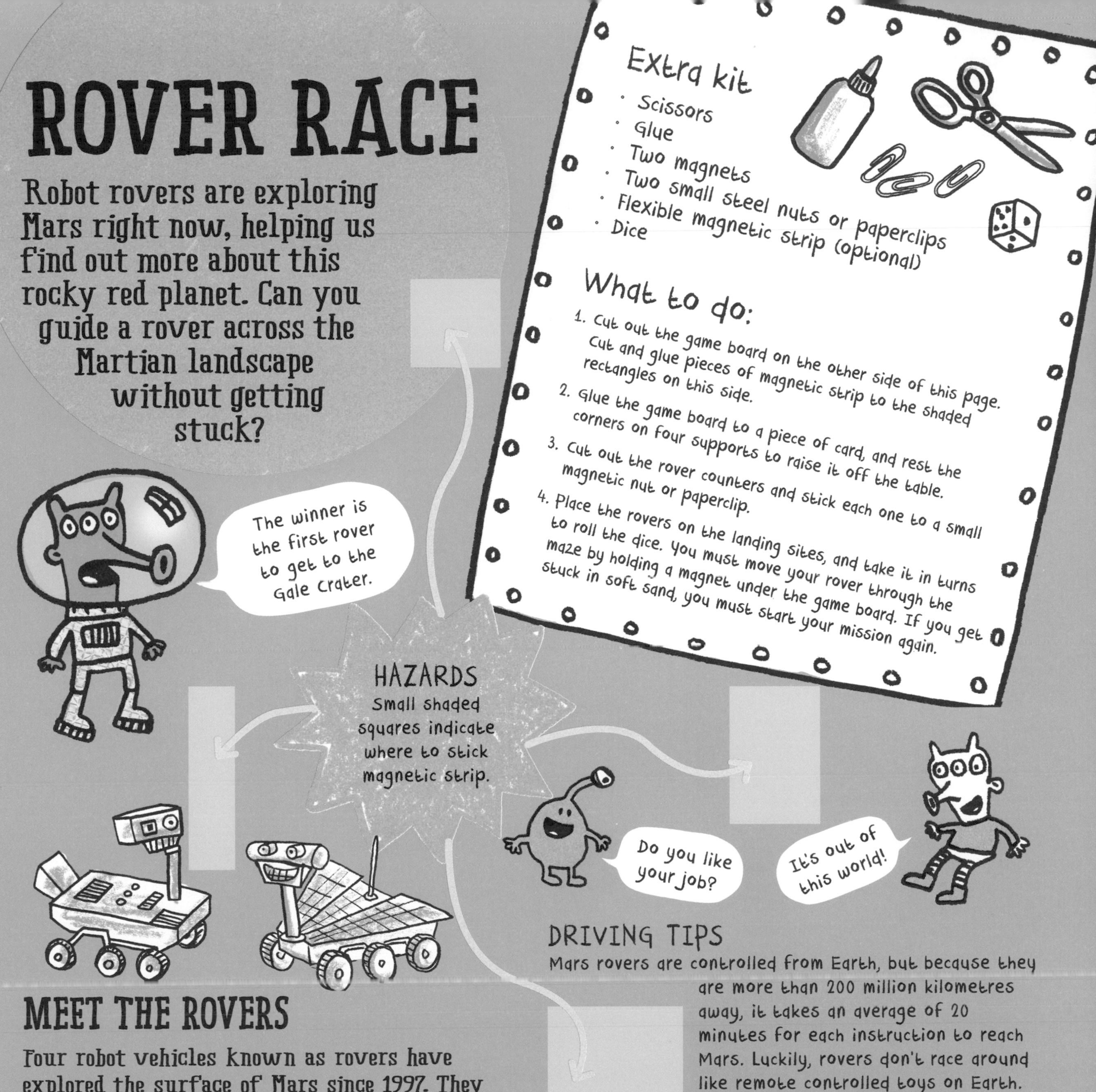


## Extra kit

- Scissors
- Glue
- Two magnets
- Two small steel nuts or paperclips
- Flexible magnetic strip (optional)
- Dice

## What to do:

1. Cut out the game board on the other side of this page. Cut and glue pieces of magnetic strip to the shaded rectangles on this side.
2. Glue the game board to a piece of card, and rest the corners on four supports to raise it off the table.
3. Cut out the rover counters and stick each one to a small magnetic nut or paperclip.
4. Place the rovers on the landing sites, and take it in turns to roll the dice. You must move your rover through the maze by holding a magnet under the game board. If you get stuck in soft sand, you must start your mission again.

### DRIVING TIPS

Mars rovers are controlled from Earth, but because they are more than 200 million kilometres away, it takes an average of 20 minutes for each instruction to reach Mars. Luckily, rovers don't race around like remote controlled toys on Earth. Curiosity rover has a top speed of around four centimetres per second!

## MEET THE ROVERS

Four robot vehicles known as rovers have explored the surface of Mars since 1997. They are like robot geologists, testing rocks and soil, taking photographs, and hunting for evidence that Mars once had all the ingredients needed to support life.

SOJOURNER

Distance travelled: 100 metres

OPPORTUNITY

Distance travelled: More than 40 kilometres

SPIRIT

Distance travelled: Almost 8 kilometres, before it got stuck in soft sand

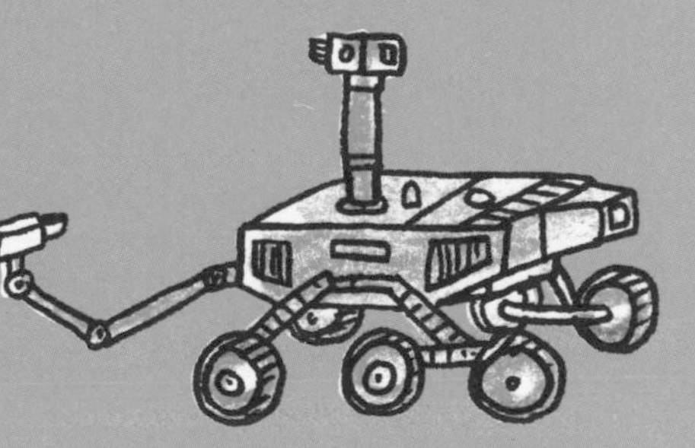

CURIOSITY

Distance travelled: More than 8 kilometres

Stop to take a rock sample – miss a turn.

Water signs detected – roll again!

Stuck in soft sand – start your mission again!

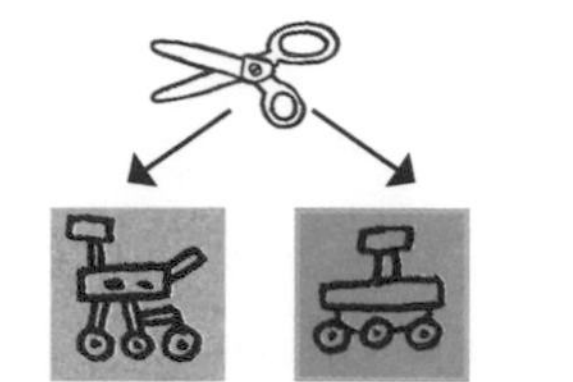

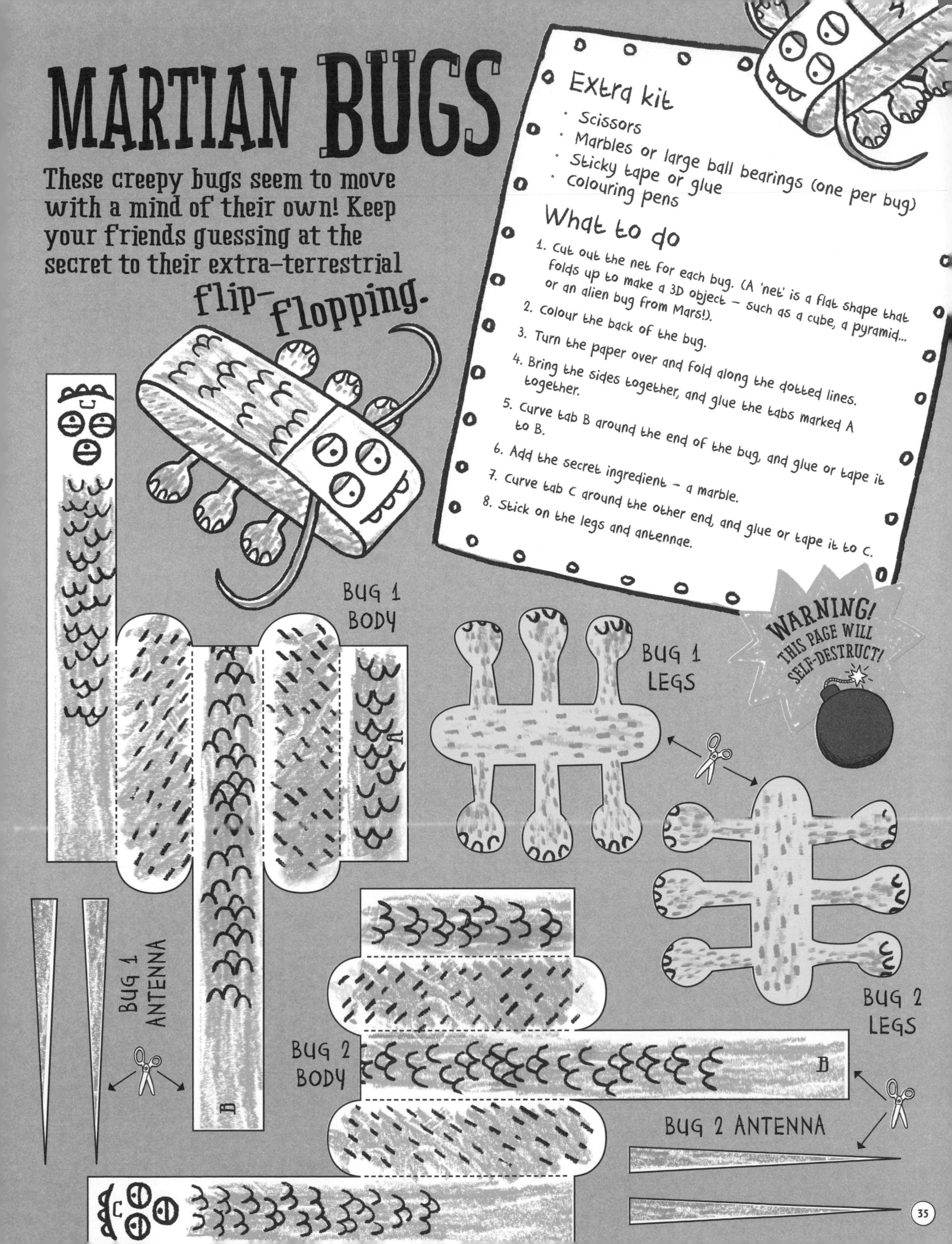

# MARTIAN BUGS

**These creepy bugs seem to move with a mind of their own! Keep your friends guessing at the secret to their extra-terrestrial flip-flopping.**

## Extra kit

- Scissors
- Marbles or large ball bearings (one per bug)
- Sticky tape or glue
- Colouring pens

## What to do

1. Cut out the net for each bug. (A 'net' is a flat shape that folds up to make a 3D object – such as a cube, a pyramid... or an alien bug from Mars!).
2. Colour the back of the bug.
3. Turn the paper over and fold along the dotted lines.
4. Bring the sides together, and glue the tabs marked A together.
5. Curve tab B around the end of the bug, and glue or tape it to B.
6. Add the secret ingredient – a marble.
7. Curve tab C around the other end, and glue or tape it to C.
8. Stick on the legs and antennae.

# WHAT'S HAPPENING?

When the bug (or the surface it's on) is tilted, the marble rolls around inside. As it crashes into the curved ends, the bug flips over – leaving the marble free to roll again. The bugs move in a random way, which makes them fun to race. It's impossible to predict which bug will win!

What do you call an alien with three eyes?

An aliiien

# ARE THERE BURPING BUGS ON MARS?

Robot rovers and space probes have been hunting for signs that microbes once lived on Mars. They have found evidence that there was once water on the red planet. This is exciting, because all living things need water. A Mars rover has also detected methane – a gas burped by living things on Earth! Does this mean there are burping bugs on Mars, or does the methane come from something else?

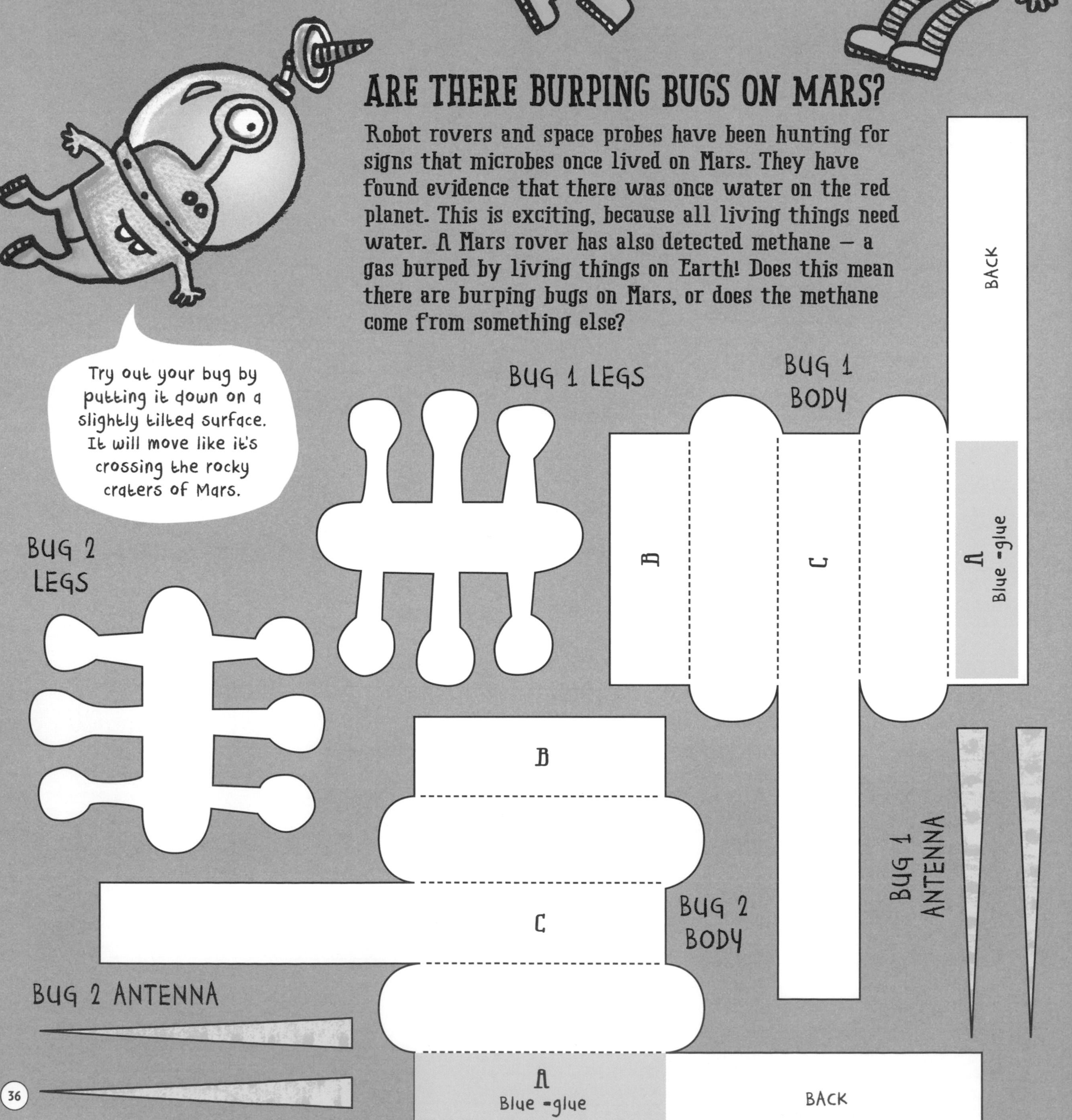

# STUCK ON YOU

Thrown off balance by Earth's tilted axis? Seasons got you in a spin? Make this flip book and watch a year go by in the blink of an eye!

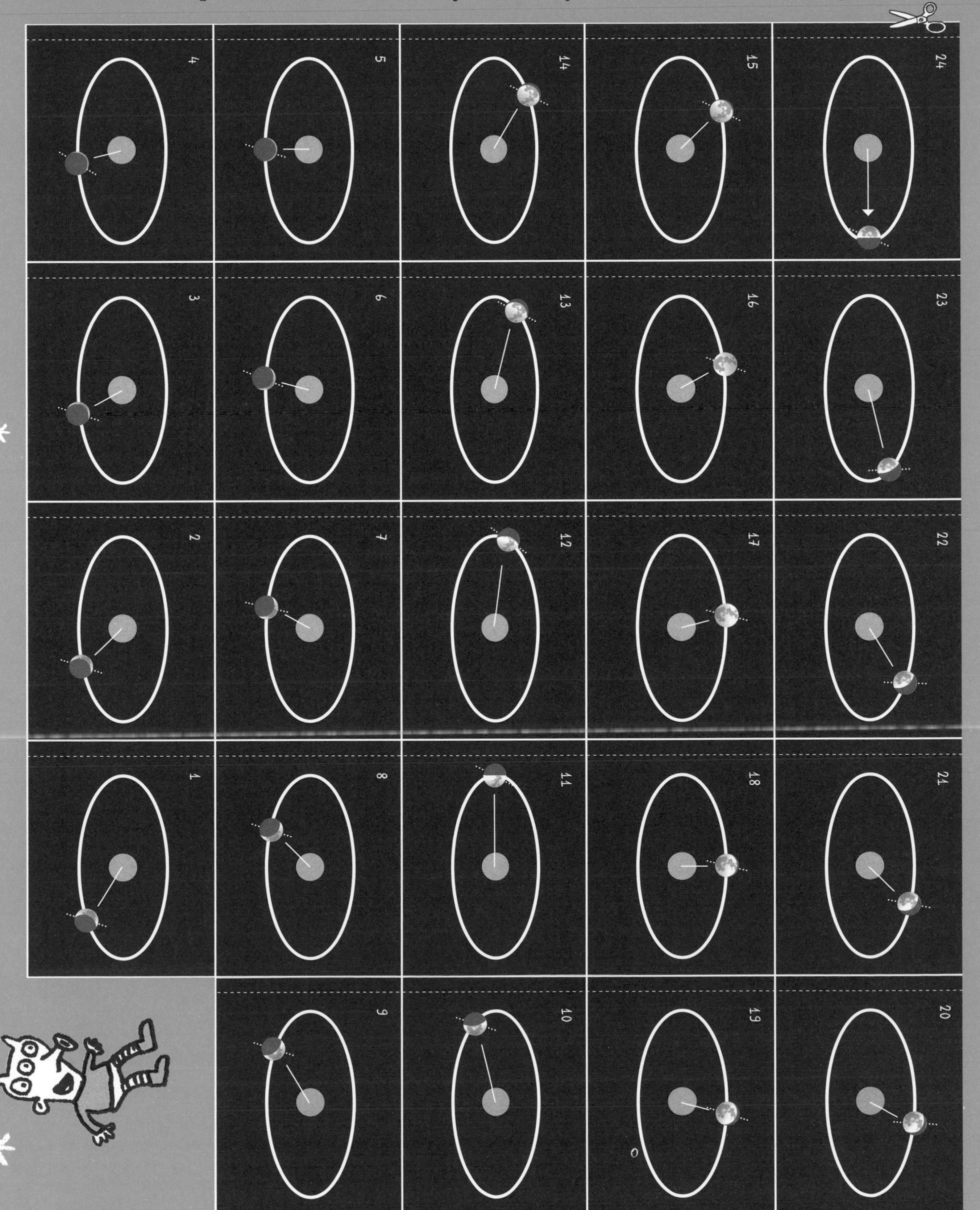

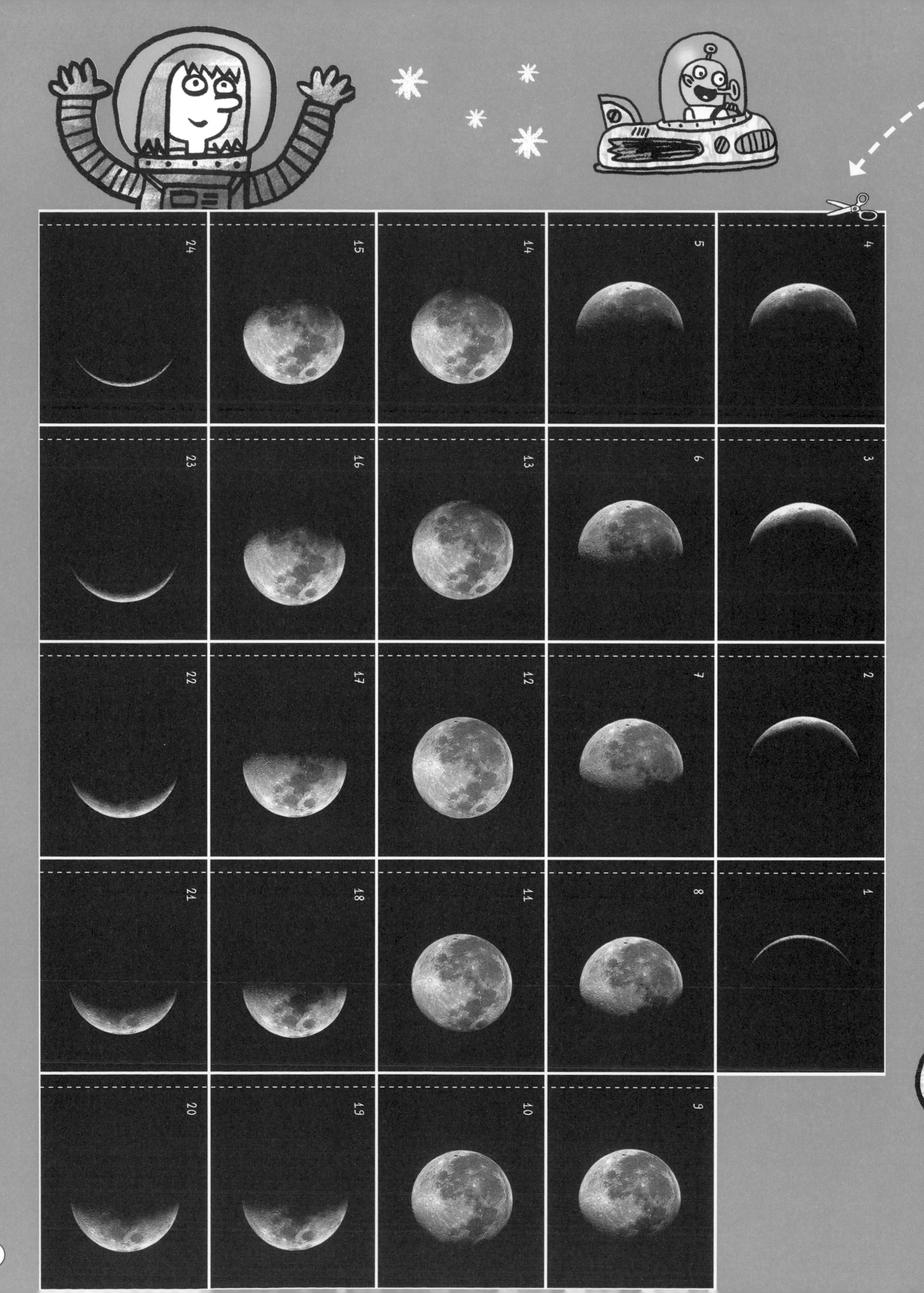
24
15
14
5
4
23
16
13
6
3
22
17
12
7
2
21
18
11
8
1
20
19
10
9

## Extra kit

- Scissors
- Stapler

## What to do

1. Cut out the 24 cards on page 38. Stack them sunny-side down from 1 to 25.
2. Staple them together at the left-hand edge on the dotted line.
3. Hold the stapled edge in your left hand and flick through the pictures with your right thumb to see Earth orbit the sun.
4. Flip the book over to watch our view of the Moon change as it orbits Earth.

Earth is tilted on its axis, meaning that each hemisphere is tilted towards the Sun for part of year (summer) and away from the Sun for part of the year (winter). The hemisphere that tilts towards the Sun gets more sunlight, so is warmer at that time of year.

A year is the time it takes for a planet to orbit its star once. Every birthday, you celebrate another 940 million-kilometre trip around the Sun! If you think 365 days is too long to wait, move to Mercury, where you'd be blowing out birthday candles every three months!

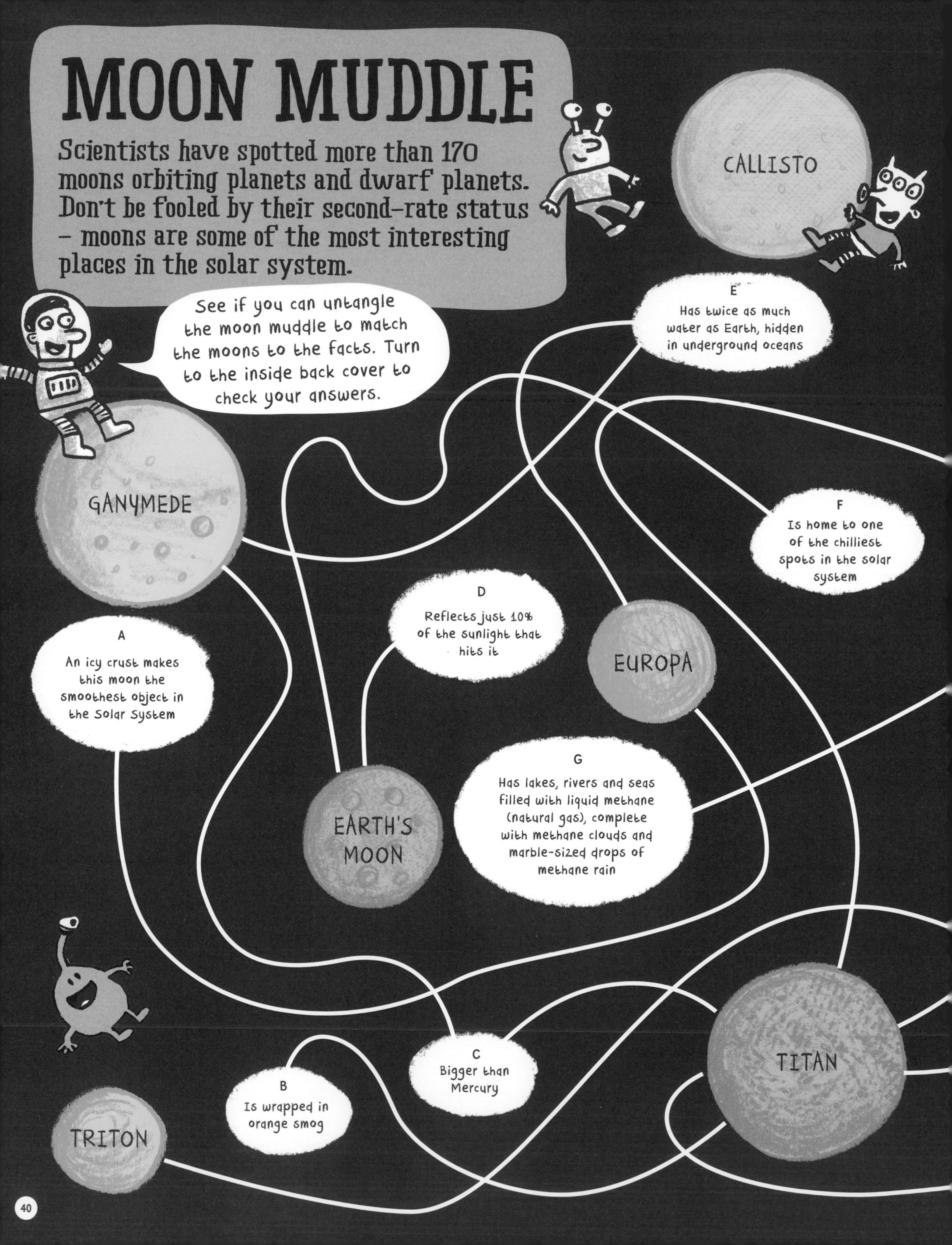

MOON MUDDLE
Scientists have spotted more than 170 moons orbiting planets and dwarf planets. Don't be fooled by their second-rate status – moons are some of the most interesting places in the solar system.
See if you can untangle the moon muddle to match the moons to the facts. Turn to the inside back cover to check your answers.
CALLISTO
E Has twice as much water as Earth, hidden in underground oceans
GANYMEDE
F Is home to one of the chilliest spots in the solar system
D Reflects just 10% of the sunlight that hits it
A An icy crust makes this moon the smoothest object in the Solar System
EUROPA
G Has lakes, rivers and seas filled with liquid methane (natural gas), complete with methane clouds and marble-sized drops of methane rain
EARTH'S MOON
C Bigger than Mercury
TITAN
B Is wrapped in orange smog
TRITON

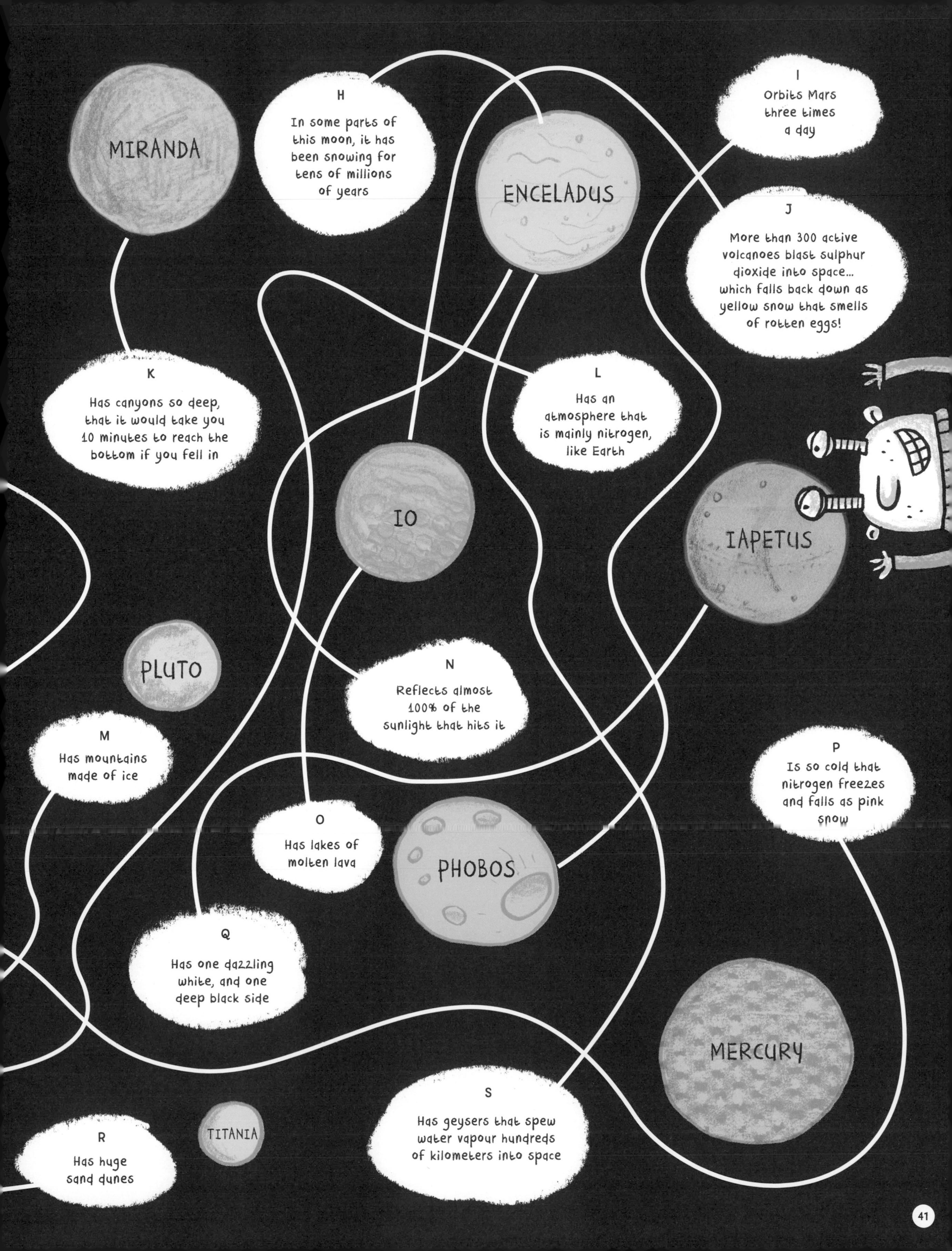
MIRANDA
H
In some parts of this moon, it has been snowing for tens of millions of years
ENCELADUS
I
Orbits Mars three times a day
J
More than 300 active volcanoes blast sulphur dioxide into space... which falls back down as yellow snow that smells of rotten eggs!
K
Has canyons so deep, that it would take you 10 minutes to reach the bottom if you fell in
L
Has an atmosphere that is mainly nitrogen, like Earth
IO
IAPETUS
PLUTO
N
Reflects almost 100% of the sunlight that hits it
M
Has mountains made of ice
P
Is so cold that nitrogen freezes and falls as pink snow
O
Has lakes of molten lava
PHOBOS
Q
Has one dazzling white, and one deep black side
MERCURY
S
Has geysers that spew water vapour hundreds of kilometers into space
R
Has huge sand dunes
TITANIA

# LUNARLYMPICS

It's time to find out how well you would run, jump and throw on the Moon. The Moon is 83 times less massive than Earth, so its gravity is much weaker. Around six times weaker!

**Extra kit:**

- Chalk
- Wall
- Football
- Tape measure
- Large garden or playing field
- Scales
- Large friend
- Calculator

## EVENT ONE – THE HIGH BOUNCE

Stand next to a wall. Hold a piece of chalk in one hand and mark how high you can reach while standing on the floor. Now jump as high as you can and mark where your hand reaches.

centimetres ········ X 6 ········> centimetres

MY BEST EARTH JUMP

MY BEST MOON JUMP

## EVENT TWO – THE LONG KICK

Mark a spot on the ground, and put the football on it. Kick it as far as you can, and measure the distance it travels.

metres ········ X 6 ········> metres

MY BEST EARTH KICK

MY BEST MOON KICK

## EVENT THREE – WEIGHTLIFTING

Ask a friend to sit on a pair of scales. Write down their starting weight in kilograms. Now take hold of their arms and pull up as hard as you can. Write down the lowest weight that appears on the scales and subtract it from the starting weight.



Replace the 6 with these numbers to calculate what would happen if the Lunarlympics was held elsewhere!

| | |
|---|---|
| MERCURY | 2.6 |
| VENUS | 1.1 |
| MARS | 2.6 |
| JUPITER | 0.4 |
| SATURN | 0.9 |
| URANUS | 1.1 |
| NEPTUNE | 0.9 |



## WHAT WOULD IT BE LIKE TO LIVE ON THE MOON?

Not a lot happens on the Moon when there are no astronauts around. There is no flowing water, and no wind to disturb the dusty surface. The Moon spins very slowly in space, with 14 Earth days between dawn and dusk. During the long nights, temperatures dip as low as –247 degrees Celsius – some of the lowest in the Solar System! There's no chance of rain but don't put that umbrella away – the Moon's thin atmosphere means that meteorites as big as golf balls rain down on the surface.

# BEDROOM PLANETARIUM

How many stars can you count on a dark night?

Millions?

Billions?

The answer is around 4,500 with your eyes alone. Grab a pair of binoculars, and this goes up to more than 200,000! To keep track, ancient astronomers used constellations.

**Extra kit:**

- Drawing pin or sharp pencil
- Old, clean cardboard tube
- Cork or foam mat
- Scissors
- Elastic band
- Torch

**What to do:**

1. Remove the opposite page from the book and lay it face down on a cork or foam mat. Carefully poke a hole through each star.
2. Cut out each circle, and cut slits around the edge. Fold the tabs over a cardboard tube. Use an elastic band to hold it in place.
3. When it's dark, shine a torch through the tube to project the constellation on to a wall or ceiling.



## STORYBOOK IN THE STARS

The constellations are 88 imaginary shapes connecting stars in the same patch of sky. Many were created by the Ancient Greeks and Romans, whose stories of the battles between monsters and heroes helped people remember the patterns. They obviously thought the skies were less peaceful than they look!

The stars that we can see from Earth change over the course of a year, so constellations were a useful way to track the seasons. They also helped astronomers share their discoveries. "Look at the end of the Great Bear's tail," is a lot quicker than saying, "check out the 701st star from the left.... Left a bit... Right a bit... Oh, I give up!"

Once you've practised spotting constellations on your ceiling, it'll be a doddle to spot them in the night sky. The stars that you see in the sky depend on where you are on Earth, and what time of year it is.

Why doesn't the Dog Star laugh when you tickle it?

It's too Sirius.

The names of the constellations are still used, but they refer to 88 patches of sky rather than the shapes themselves.

CENTAURUS / THE CENTAUR

In Greek myths, the Centaur had the body of a man, and the legs of a horse.

ANDROMEDA

Andromeda was a princess who was chained to a rock as a sea monster's supper, but rescued.

URSA MAJOR / THE GREAT BEAR

Seven of its stars make up a famous pattern called the Plough.

ORION / THE HUNTER

Orion was a great hunter in Greek myths. The line of three stars in the middle is called Orion's belt.

LEO / THE LION

This was the lion that Hercules wrestled.

HERCULES

Hercules was a super-strong Greek hero who battled with many beasts. In his constellation, he is standing on the head of a dragon!

CETUS / THE SEA MONSTER

This was the sea monster sent to eat Andromeda.

SCORPIUS / THE SCORPION

After Orion boasted that he could kill any wild animal, this Scorpion was sent to finish him off!

PEGASUS / THE WINGED HORSE

Pegasus was a winged horse that helped a Greek hero to kill a fire-breathing monster.

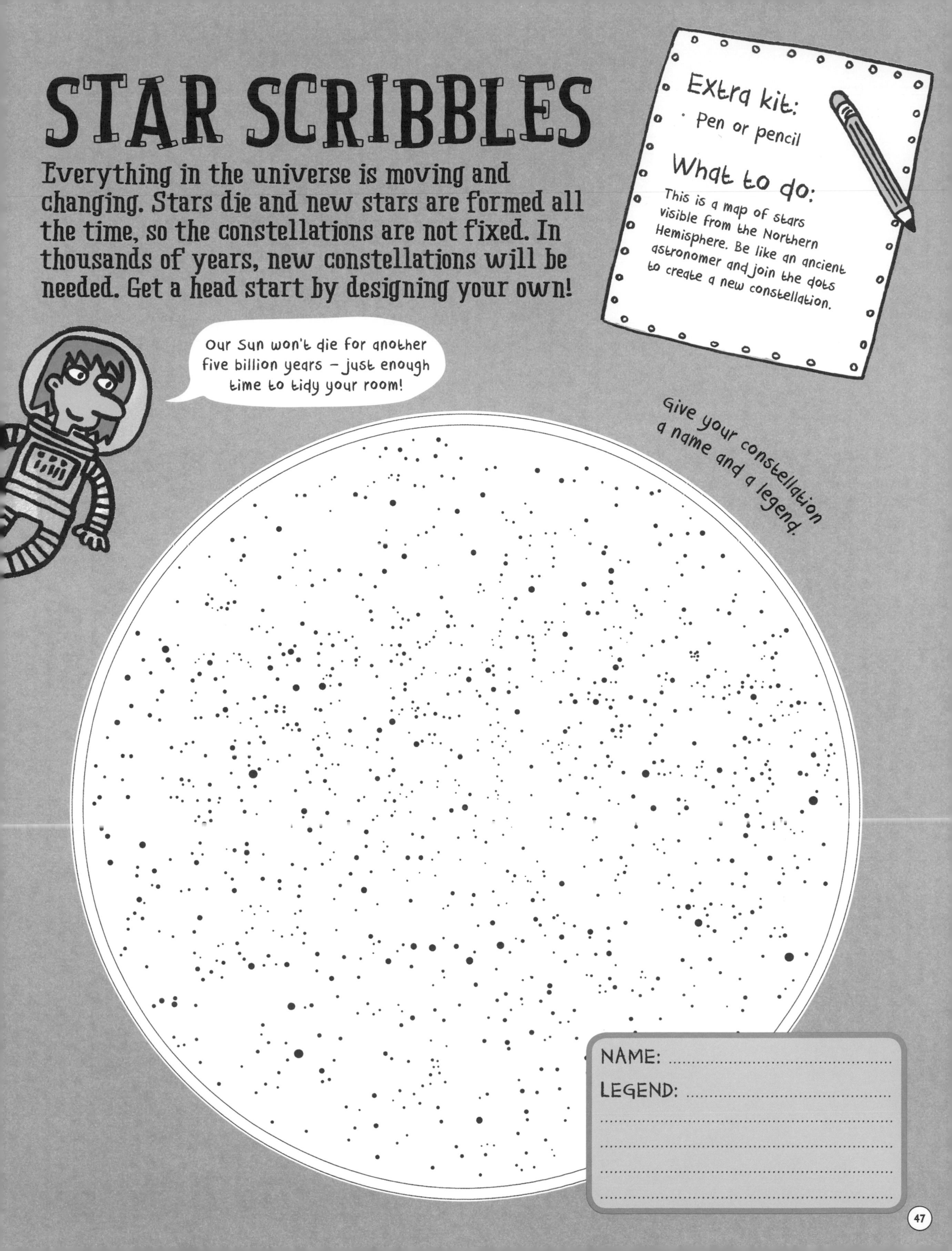
STAR SCRIBBLES
Everything in the universe is moving and changing. Stars die and new stars are formed all the time, so the constellations are not fixed. In thousands of years, new constellations will be needed. Get a head start by designing your own!
Extra kit:
· Pen or pencil
What to do:
This is a map of stars visible from the Northern Hemisphere. Be like an ancient astronomer and join the dots to create a new constellation.
Our Sun won't die for another five billion years – just enough time to tidy your room!
Give your constellation a name and a legend.
NAME: ...................................
LEGEND: ...................................

If you stand in a different place on Earth, you'll see a different set of constellations! This is a map of stars visible from the Southern Hemisphere.

Give your constellation a name and a legend.

NAME: ..........................................

LEGEND: ..........................................

..........................................

..........................................

..........................................

..........................................

What's an astronaut's favourite key on a keyboard?

The space bar!

# GALAXY GENERATOR

A galaxy is a huge, spinning group of stars, dust and gas, held together by gravity. Grab some scissors and glue, and turn this page into a beautiful spiral galaxy!

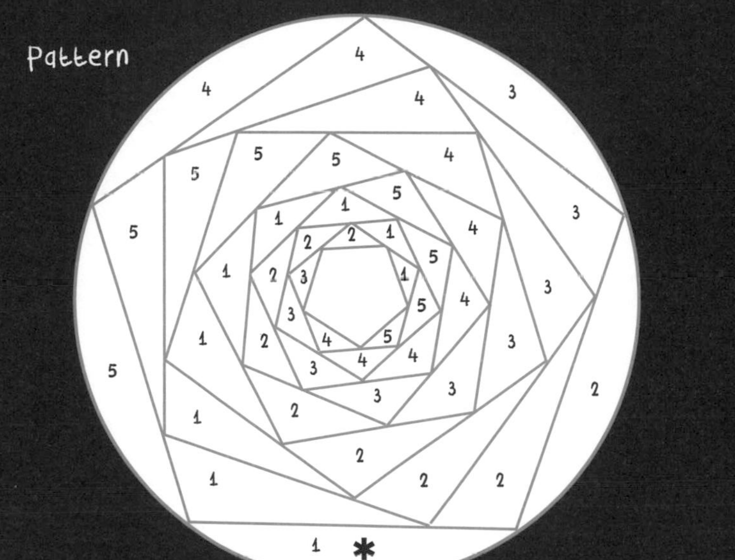

## Extra kit:

- Scissors
- Glue
- Sticky tape
- Small piece of silver foil

## What to do

1. Cut out the base, and carefully cut the hole from the centre.
2. Put the base face down over the pattern and hold it in place with masking tape.
3. Cut out the 40 small strips of paper, keeping them in numbered piles.
4. Dab some glue onto the blue area of the base, and stick the first strip of paper in place over the area of the pattern marked with a 1 and a *. Work your way around the base anti-clockwise, placing strips from each group over the right numbers on the pattern. Make sure you don't glue the strips to the pattern – just the base.
5. Keep adding strips until you reach the middle. Glue the piece of foil over the hole.
6. Cover the back of the strips with sticky tape. Finally, cut around the dashed line to trim the spiral galaxy into a circular shape.



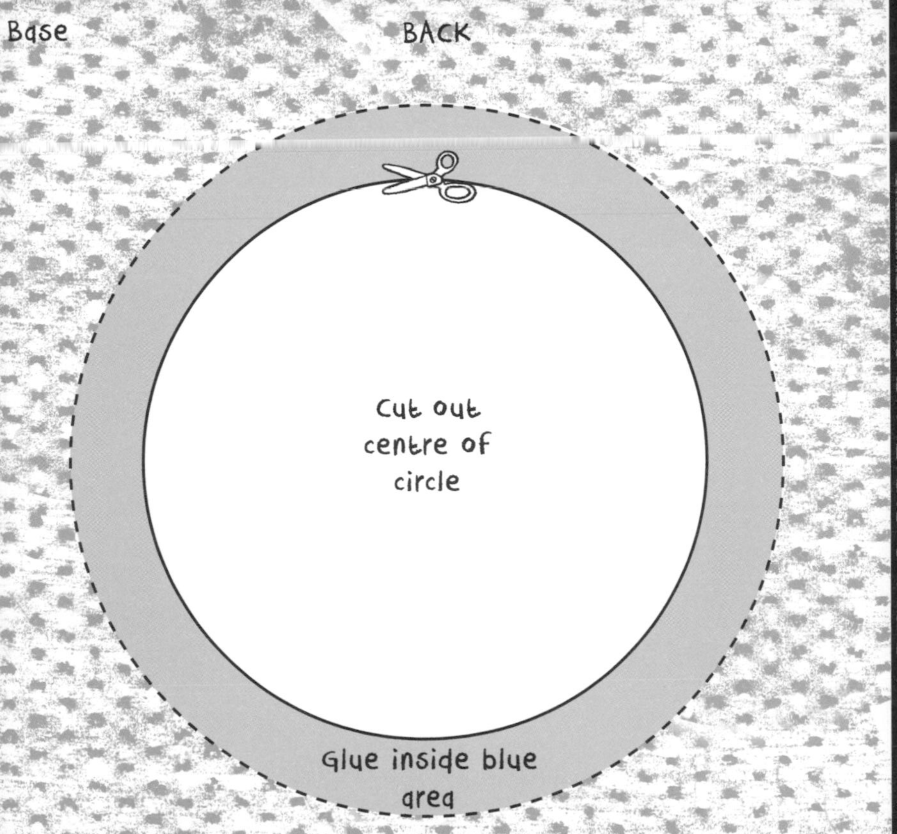

Almost three-quarters of the galaxies that can be seen from Earth are spiral galaxies, with long arms spiraling out from a bright bulge at the centre. Our own galaxy, the Milky Way, is a spiral galaxy.

Our Sun takes around 250 million years to make one orbit of the Milky Way, taking Earth and the other planets along for the ride. The last time Earth was this side of the galaxy, it was dinosaurs, not humans, enjoying the view!

# STEP RIGHT THROUGH

Have you ever wished you could leap through a wormhole into another world, like they do in science fiction films? Warm up by learning to leap through this page!

WARNING! THIS PAGE WILL SELF-DESTRUCT!

*You might not find another world on the other side, but you will amaze your friends.

Extra kit:

Scissors

What to do:

1. Start by tearing this page out of the book. Fold it in half along the dotted black line.
2. Cut along the solid black lines - do not cut all the way to the edge.
3. You now have lots of mini folds. Leave the first and last folds intact, but cut through the others to make a big hole.
4. Open out the hole and step right through!

It's one small step...

DON'T CUT THIS FOLD

Cut the dotted folds between the stars.

Cut the dotted folds between the stars.

DON'T CUT THIS FOLD

# WHAT'S A WORMHOLE?

It's a shortcut through space and time – in theory! Space scientists think wormholes could be formed when space gets bent or distorted, bringing two distant points together (a bit like bringing opposite ends of this page together by bending the paper in half). Wormholes sound exciting because they could make it possible to travel huge distances and explore different solar systems. But scientists haven't yet found proof that they really exist.

SLURP!

SPIT!

Einstein came up with a theory about space.

About time too.

Once you've learned how, try the trick with a smaller piece of paper. Can you leap through a birthday card or a bus ticket?

Challenge your friends to see who can fit the most people inside the perimeter of a tiny piece of paper. Use this trick to win every time!

# MIND THE BLACK HOLE!

Watch out – you're being sucked towards a black hole at the centre of a spiral galaxy.

Will you be stretched into space-spaghetti, or burnt to a crisp?

WARNING! THIS PAGE WILL SELF-DESTRUCT!

## Extra kit

- Scissors
- Thin cardboard
- Two one centimetre counters
- Dice
- A fellow astronaut

## What to do

1. Take this page out of the book and stick it to a piece of thin cardboard with the game board face up. Remember to read the rest of this side first!
2. Carefully cut out the six wormholes. Balance the cardboard on top of a box, so that counters can fall through the wormholes.
3. Follow the instructions on the game board to play.

Scientists think there is a supermassive black hole at the centre of most galaxies – including the Milky Way. The bigger the bulge at the centre of a galaxy, the bigger the black hole! Black holes are strange areas of space where an enormous gravitational pull squeezes matter into a tiny area. The gravitational pull is so strong, not even light can escape. Like whirlpools, swirling discs of space dust and stars surround black holes.

## CRISP OR SPAGHETTI?

Scientists can't see black holes – just their effects on nearby objects. No one is completely sure what happens to objects that get sucked in. An astronaut who got too close might be stretched into spaghetti by the gravitational pull, or burned to a crisp by a superheated 'firewall' of energy. Either way, you wouldn't have long to worry about it – you'd be crushed once you got to the core!

It's hard to choose!

## How to play

Two players

1. Put your counters on the starting positions at the edges of the galaxy.
2. Take it in turns to roll the dice and move.
3. If you land on a symbol, follow the instructions.
4. The winner is the astronaut who avoids being sucked into the black hole!

Dazzled by a supernova, miss a turn

Fallen into a wormhole, start again

Gravity boost, move the number of spaces shown

CUT OUT BEFORE PLAY

# UNIDENTIFIED FLYING OBJECTS

## Extra kit

- Scissors
- Sticky tape

UFO stands for Unidentified Flying Object. People often think they are alien spacecraft, but many 'sightings' have actually been hoaxes. Make your own UFO and see if you can take photographs that fool your family!

WARNING! THIS PAGE WILL SELF-DESTRUCT!

Earth is home to the only life in the universe... that we know about. But the Universe is so humungous, space scientists think there must be LOADS more planets out there with conditions perfect for life. No wonder everyone gets so excited when they spot a UFO!

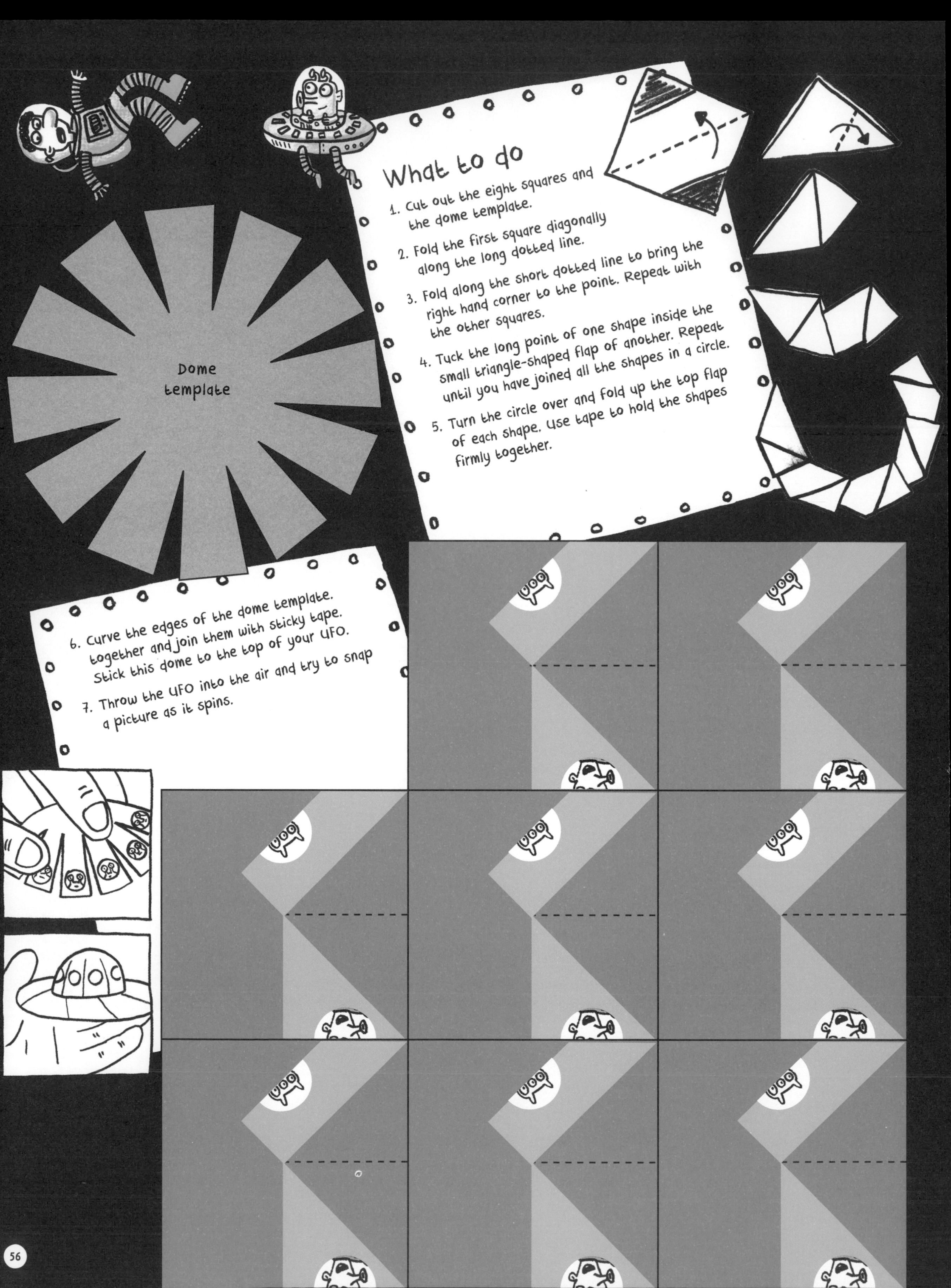
What to do
1. Cut out the eight squares and the dome template.
2. Fold the first square diagonally along the long dotted line.
3. Fold along the short dotted line to bring the right hand corner to the point. Repeat with the other squares.
4. Tuck the long point of one shape inside the small triangle-shaped flap of another. Repeat until you have joined all the shapes in a circle.
5. Turn the circle over and fold up the top flap of each shape. Use tape to hold the shapes firmly together.
Dome template
6. Curve the edges of the dome template together and join them with sticky tape. Stick this dome to the top of your UFO.
7. Throw the UFO into the air and try to snap a picture as it spins.

# ALIEN ID

**Aliens are out there – but what do they look like? Creatures living in Earth's most extreme environments could give us clues.**

## POMPEII WORM

These worms are the most heat-tolerant animals on Earth. They live right next to hydrothermal vents on the sea floor, and are often spotted with their tails dangling in 80 degrees Celsius water – do NOT try this at home.

## BLOBFISH

Blobfish live in deep oceans, where the pressure is up to 120 times greater than it is at the surface. Its jelly-like blob of a body helps it float above the sea floor without being crushed.

## HIMALAYAN JUMPING SPIDER

These tiny spiders live in mountains 6,700 metres above sea level, higher than almost any other animal in the world. This means they can cope with very low oxygen levels.

## YETI CRAB

Yeti crabs don't have eyes – there's no point in the darkness of the world's deepest oceans. Instead of hunting for food, they grow their own 'gardens' of bacteria on their bristly claws.

## DEEP-SEA AMPHIPOD

These super-sized shrimps live in deep ocean trenches, where the pressure would crush a human. Tough stomachs help them digest junk that sinks to the bottom, including wood from shipwrecks.

## COCKROACH

These insects are so good at adapting to different conditions, they have been on Earth for more than 300 million years. They can survive huge doses of radiation, one of the biggest hazards in space.

## SPINOLORICUS CINZIA

These mysterious creatures survive in a 'dead zone' at the bottom of an ocean. Unlike every other animal on Earth, they live without any oxygen at all.

## GIANT TUBE WORM

These two-metre-long worms live near hydrothermal vents, where water heated by molten rock gushes out of the ocean floor. This water is full of chemicals that would kill most living things.

## SAHARA SILVER ANT

These ants survive in the world's hottest desert, sprinting out of their burrows to take advantage of an all-you-can-eat buffet of insects that have been killed by the extreme heat!

## TARDIGRADE

These incredible creatures have come back to life after being heated above 150 degrees Celsius and below -270 degrees Celsius. Tardigrades were sent to space for ten days – without space suits – and were still happily going about their business when they came back to Earth!

## SEA PIG

This animal is actually a type of sea cucumber! It feasts on mud and dead animals at the bottom of the world's deepest area of ocean, the Mariana Trench.

**All these creatures live in conditions that would kill most living things – conditions like those on other planets and moons of the Solar System. Scientists call them extremophiles.**

Design your own alien life form, using features from Earth's extremophiles!
Alien ID:
Home planet or moon:
Habitat:

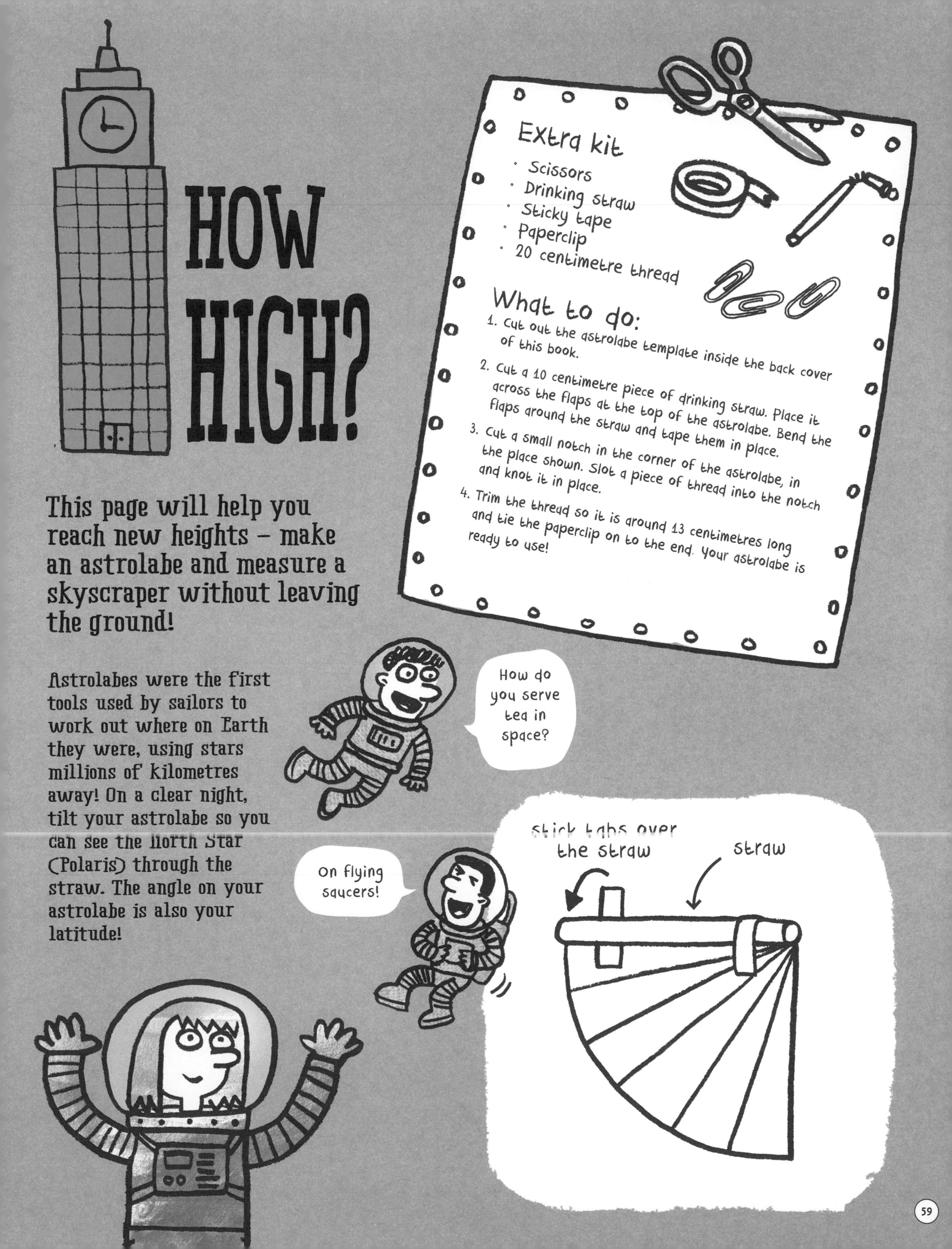
HOW HIGH?
This page will help you reach new heights – make an astrolabe and measure a skyscraper without leaving the ground!
Astrolabes were the first tools used by sailors to work out where on Earth they were, using stars millions of kilometres away! On a clear night, tilt your astrolabe so you can see the North Star (Polaris) through the straw. The angle on your astrolabe is also your latitude!
Extra kit
• Scissors
• Drinking straw
• Sticky tape
• Paperclip
• 20 centimetre thread
What to do:
1. Cut out the astrolabe template inside the back cover of this book.
2. Cut a 10 centimetre piece of drinking straw. Place it across the flaps at the top of the astrolabe. Bend the flaps around the straw and tape them in place.
3. Cut a small notch in the corner of the astrolabe, in the place shown. Slot a piece of thread into the notch and knot it in place.
4. Trim the thread so it is around 13 centimetres long and tie the paperclip on to the end. Your astrolabe is ready to use!
How do you serve tea in space?
On flying saucers!
stick tabs over the straw
straw
59

1. Line up the straw so you can see the top of the building through it. When the thread comes to rest, hold it against the card and read the angle.

2. Find your measured angle in the table, and write the number next to it in this box: □

3. Now grab a calculator and do this speedy sum:

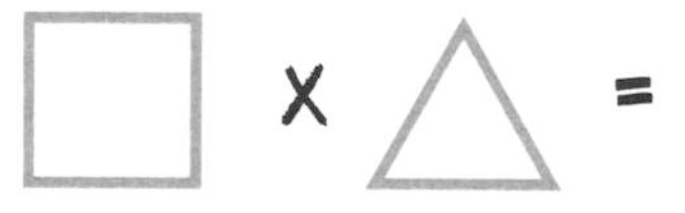

4. Add the number in the ○ (your height in metres) to your answer to reveal the height of the building! Now, where's that ladder...

# TANGENT TABLE

| ANGLE | | ANGLE | |
|---|---|---|---|
| 5 | 0.0875 | 50 | 1.1918 |
| 10 | 0.1763 | 55 | 1.4281 |
| 15 | 0.2679 | 60 | 1.7321 |
| 20 | 0.3640 | 65 | 2.1445 |
| 25 | 0.4663 | 70 | 2.7475 |
| 30 | 0.5773 | 75 | 3.7321 |
| 35 | 0.7002 | 80 | 5.6713 |
| 40 | 0.8391 | 85 | 11.430 |
| 45 | 1 | | |

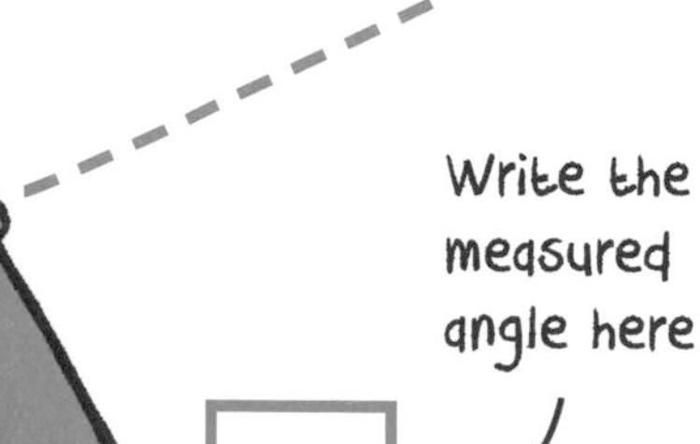

Write the measured angle here

Height

Write the height from the ground to your eye here

Distance

Write the distance from your feet to the building here

You can use your astrolabe to measure the height of anything – you just need to know how far you are from the object. A map can help you find out.

# SPACE JUNK

How long can you stay in orbit around Earth without

crashing into space junk,

or bashing into

an asteroid?

## Extra kit

- Counters
- Dice
- Two or more players

## What to do

1. Put the game board on a flat surface. Launch your counter from Earth by rolling the dice and following the numbered path.
2. Once you are in orbit, move from place to place by rolling the dice and following the numbered path.
3. If you crash into a piece of space junk, you must roll again to avoid it. If you land on Earth, the Moon or the International Space Station, you can miss a go while resting in a safe place. If you crash into an asteroid you are out!
4. The winner is the last player left in orbit.

## SUPER-SONIC SCRAP

Millions of pieces of space junk are whizzing around Earth right now. They include old satellites, used-up bits of rocket, debris from crashes in space, and anything that astronauts have accidentally dropped during spacewalks. Like other orbiting objects, space junk travels at up to 28,000 kilometres per hour – more than 22 times the speed of sound! At that speed, even a small speck of paint can damage a spacecraft or satellite. To keep astronauts and the International Space Station safe, NASA tracks half a million pieces of space junk that are larger than marbles.

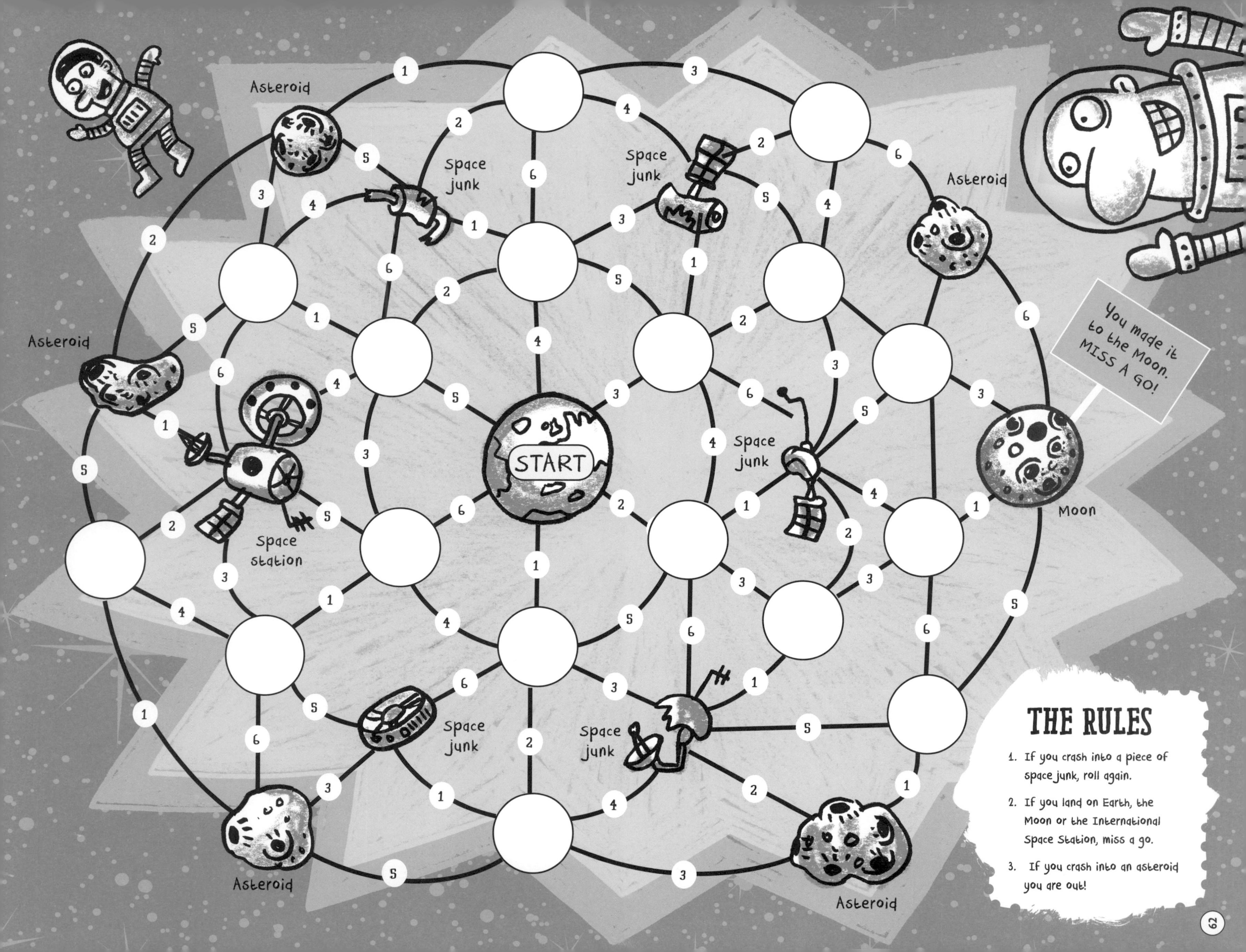


## THE RULES

1. If you crash into a piece of space junk, roll again.
2. If you land on Earth, the Moon or the International Space Station, miss a go.
3. If you crash into an asteroid you are out!

# SPACED OUT

This page morphs into a 'Don't Disturb' sign for guaranteed peace and quiet while you're busy destroying this book!

## Extra kit

- Scissors
- Two photographs of you, trimmed to 6 x 6 centimetres

## what to do

1. Cut out the template, including the marked shapes.
2. Glue your photographs face down in the places shown.
3. Fold along the dotted line and glue the flaps together.

WARNING! THIS PAGE WILL SELF-DESTRUCT!

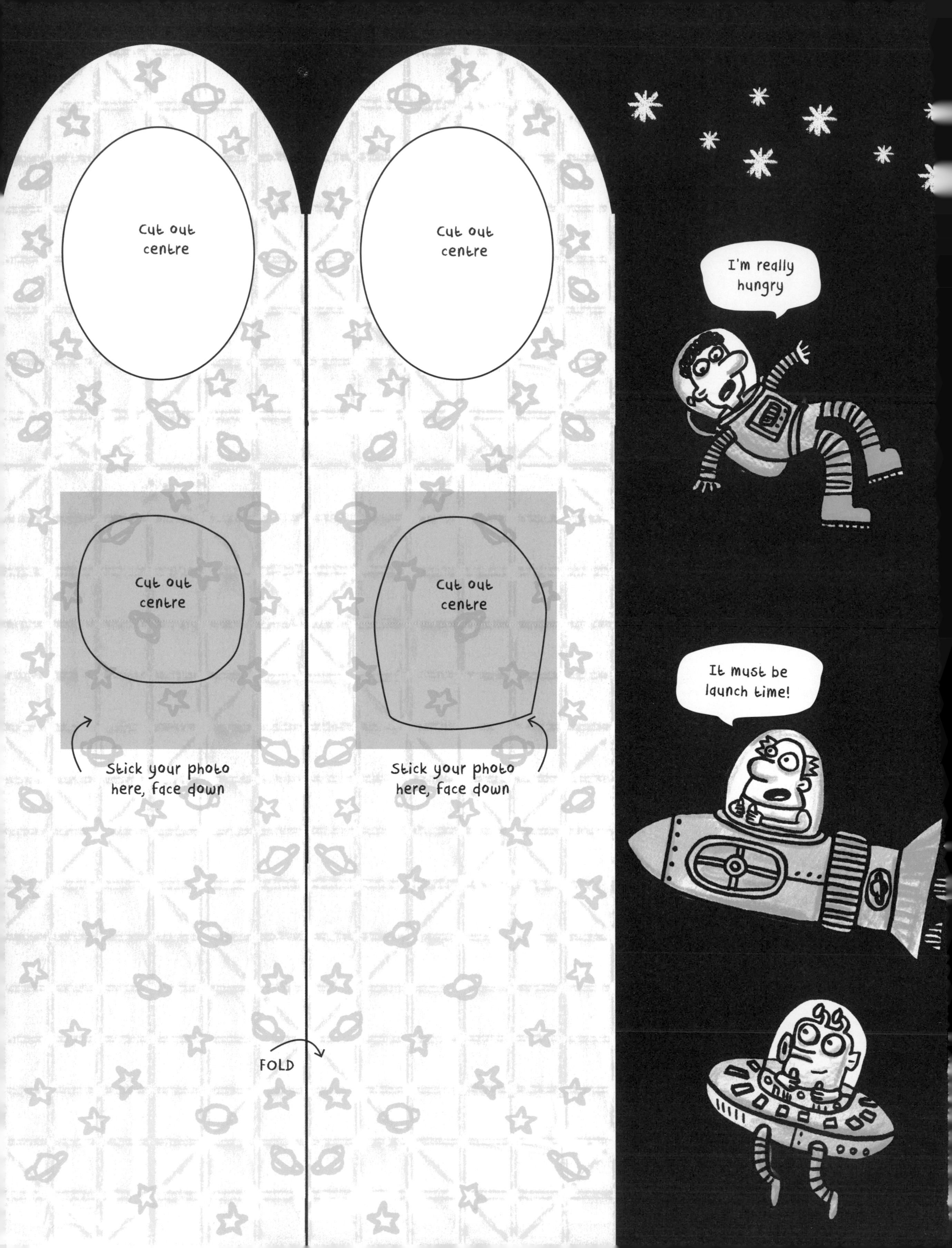
Cut out centre
Cut out centre
Cut out centre
Cut out centre
Stick your photo here, face down
Stick your photo here, face down
FOLD
I'm really hungry
It must be launch time!